NUCLEAR ENERGY

WHAT EVERYONE NEEDS TO KNOW®

NUCLEAR ENERGY

WHAT EVERYONE NEEDS TO KNOW®

CHARLES D. FERGUSON

OXFORD
UNIVERSITY PRESS

OXFORD
UNIVERSITY PRESS

Oxford University Press, Inc., publishes works that further
Oxford University's objective of excellence
in research, scholarship, and education.

Oxford   New York
Auckland   Cape Town   Dar es Salaam   Hong Kong   Karachi
Kuala Lumpur   Madrid   Melbourne   Mexico City   Nairobi
New Delhi   Shanghai   Taipei   Toronto

With offices in
Argentina   Austria   Brazil   Chile   Czech Republic   France   Greece
Guatemala   Hungary   Italy   Japan   Poland   Portugal   Singapore
South Korea   Switzerland   Thailand   Turkey   Ukraine   Vietnam

Published by Oxford University Press, Inc.
198 Madison Avenue, New York, New York 10016

www.oup.com

Oxford is a registered trademark of Oxford University Press

Library of Congress Cataloging-in-Publication Data
Ferguson, Charles D.
Nuclear energy / Charles D. Ferguson.
p. cm. — (What everyone needs to know)
ISBN 978-0-19-975945-3 (hardback) — ISBN 978-0-19-975946-0 (pbk.) 1. Nuclear energy.
I. Title. II. Series.
TK9145.F47   2011
333.792'4—dc22   2010044449

For Catharine,
who started the conversation on nuclear energy

CONTENTS

PREFACE   xiii
ACKNOWLEDGMENTS   xv

1   Fundamentals   3

What is energy, and what is power?   3
Is most energy that people use derived from
nuclear energy?   5
What is the origin of nuclear energy?   7
What is "radioactivity"?   12
Why is ionizing radiation a health concern?   13
What is radioactive half-life, and how can
knowing it help increase a country's security?   15
How much more energetic is nuclear energy
compared to chemical energy?   16
Why has it been so difficult to commercialize
nuclear fusion?   17
How was nuclear fission discovered?   19
What role, if any, did Albert Einstein play in the
discovery of nuclear energy?   21
What is a fission chain reaction?   22
What is uranium, where did it come from, and
how was it discovered?   23
What is plutonium, how was it discovered, and
how hazardous is it?   26

Why can't nuclear reactors explode like a nuclear bomb?   28

What is the nuclear fuel cycle?   28

*Why are certain activities in the nuclear fuel cycle called
"dual use"?*   32

What are the various uranium-enrichment methods?   33

*What are the nuclear-proliferation concerns for uranium
enrichment?*   36

What is the thorium fuel cycle?   38

How does a nuclear reactor generate electricity?   40

*How many people's electricity demands can be supported
with one large nuclear reactor?*   41

*What are the different types of nuclear reactors used for
electricity generation?*   41

*Why were only a few types of reactor designs chosen for
the present fleet of reactors?*   47

*What are the Generation IV reactors, and why are they
considered potentially revolutionary?*   48

What can nuclear reactors do besides generate electricity?   51

2   Energy Security and Costs of Building
Power Plants   53

What is energy security?   53

Is energy independence feasible?   55

Have countries ever been shut out of the nuclear-fuel market?   57

*Are European countries too dependent on Russian
energy supplies?*   58

*What role has nuclear energy played in reducing certain
countries' dependence on fossil fuels?*   60

*How could nuclear energy further reduce dependence on
fossil fuels?*   62

*What countries use commercial nuclear power, and how
much electricity do they obtain from it?*   63

*How many more countries are likely to acquire commercial
nuclear power plants?*   64

*How do the costs of nuclear plants compare to other types of
power plants?*   68

How can nuclear power plants be made more
cost-competitive?   70

Why is it difficult for the supply chain to keep up with
forecasts of demand for new nuclear plants?   74

How many skilled people are required to build and operate
nuclear plants?   76

Can construction of nuclear power plants keep pace with
the increasing demands for electricity?   77

Will the world run out of uranium and, if so, when?   78

Why have some countries pursued reprocessing of spent
nuclear fuel for commercial purposes?   80

Why did the United States decide to not pursue reprocessing,
and will it revive this practice?   83

3   Climate Change   86

What is the greenhouse effect?   86

What is the difference between global warming and
climate change?   88

What are the observed and forecasted effects of
climate change?   89

What can people do to reverse excess global warming?   91

Why don't nuclear plants emit greenhouse gases?   93

Why does the nuclear fuel cycle emit some greenhouse gases?   94

How helpful have nuclear power plants been in
preventing more greenhouse gases from being emitted
from coal and natural gas plants?   95

How many additional nuclear plants would be needed to
make a further major reduction in greenhouse gas emissions?   95

Have nuclear power plants ever been built as fast as would
be needed to make another major reduction in greenhouse
gas emissions?   97

Will global warming actually reduce the ability to use
nuclear power plants to their full capacity?   98

Should nuclear power be considered a "clean" energy source
for climate-change agreements among nations?   98

What are the differing views among environmentalists on
nuclear power?   100

4   Proliferation   103

What is nuclear proliferation?   103
Which countries have developed nuclear weapons,
and how did they do it?   104
How many nuclear weapons do the nuclear-armed
countries have?   108
How much weapons-usable fissile material is available
worldwide, and where is it located?   110
Has a country ever completely dismantled or given up
its nuclear arsenal?   113
What is the nonproliferation regime?   114
What is the nuclear Non-Proliferation Treaty?   114
What is the International Atomic Energy Agency,
and what role does it play in preventing proliferation?   118
What are nuclear safeguards, and how have they evolved?   119
Are nuclear safeguards effective?   123
What is the Nuclear Suppliers Group?   125
Has commercial nuclear power ever been used to make
nuclear weapons?   128
What can be done to prevent proliferation?   129
Can the nuclear fuel cycle be made more
proliferation resistant?   131
Can terrorists make nuclear weapons?   132
What can be done to prevent nuclear terrorism?   135

5   Safety   137

What is nuclear safety?   137
How safe is safe enough, and what is safety culture?   137
What is the defense-in-depth safety concept?   138
What are the major types of nuclear accidents?   140
How is nuclear safety measured?   141
How safe are today's nuclear power plants?   142
Should a country choose one plant design instead of
multiple designs, and what are the implications of such
a choice?   144
What is the China syndrome?   146

How did the Three Mile Island accident happen and
what were the consequences? 147

How did the Chernobyl accident happen and what were
the consequences? 149

What happened to Soviet-designed reactors after the
Chernobyl accident? 153

How did the nuclear industry form self-policing
organizations? 156

Can the nuclear industry survive if another major
accident occurs? 157

How long can nuclear power plants operate? 158

How can future nuclear power plants be made safer? 159

Can nuclear power expand too fast to keep plants safe? 161

How can nuclear facilities be made resistant to natural
disasters such as earthquakes and tsunamis? 163

How did the Fukushima Daiichi accident happen? 163

Were there design flaws in the Fukushima Daiichi reactors
and were there safety concerns prior to the accident? 167

What are the concerns about nuclear safety culture in Japan? 169

What are the likely implications for the nuclear industry
as a result of the Fukushima Daiichi accident? 170

Why did the Fukushima Daiichi accident raise renewed
concerns about the use of plutonium in nuclear fuel? 171

6 Physical Security 173

What is nuclear security? 173

What is a design-basis threat assessment? 174

How are safety and security different? 175

Why would someone attack a nuclear power plant or
related nuclear facility? 176

Who would attack nuclear facilities? 176

What are the potential modes of attack or sabotage,
and what has been done to protect against them? 181

What more can be done to strengthen the security of
existing and future facilities? 185

What military attacks have there been on nuclear reactors? 186

What can countries do to protect their nuclear facilities
from military attacks? 188

7   **Radioactive Waste Management**   **189**

What are the types of radioactive waste, and how are
they generated?   189

What is the typical composition of spent nuclear fuel?   190

How long does the radioactivity in spent fuel last?   191

How hazardous is radioactive waste?   192

How much spent nuclear fuel has been produced?   193

What are the storage options for dealing with
radioactive waste?   193

What is the volume of radioactive waste, and how does
this compare to other industrial toxic waste?   194

How vulnerable are spent fuel pools?   195

Does reprocessing reduce the amount of nuclear waste?   196

How do nuclear plants and coal plants compare in
terms of radioactivity emitted?   197

What country is closest to opening a permanent nuclear waste
repository?   198

How hazardous is the transportation of radioactive waste?   198

Why was Yucca Mountain chosen as the permanent
repository in the United States, and what will be its fate?   200

Will delays in opening a permanent repository for radioactive
waste derail continued or expanded use of nuclear power
in the United States?   201

8   **Sustainable Energy**   **203**

What is meant by a "sustainable energy system"?   203

What is a "renewable energy source"?   204

Is nuclear energy a renewable energy source?   205

Can nuclear energy contribute to developing sustainable
energy systems?   205

Can renewable energies compete with nuclear and other
base-load electrical power sources?   206

SUGGESTIONS FOR FURTHER READING   **209**

INDEX   **213**

PREFACE

Why should people care and learn about nuclear energy? The world faces growing concerns about a changing climate that most likely has come about largely from increasing levels of greenhouse gas emissions. While no single energy technology will solve the problem of climate change, nuclear energy has a significant role to play because it produces very low greenhouse gas emissions. Though today, nuclear power generates only a relatively small fraction of the world's electricity, it offers the potential for providing a tremendous amount of commercial electrical power as long as its production is cost-competitive. In addition, nuclear energy can buffer people's increasing concerns about the security of the world's energy supplies. By diversifying the mix of electricity-generation sources, we reduce our reliance on vulnerable fuel sources.

These benefits are substantial, but nuclear energy does present risks. Although safety at plants has improved significantly since the 1979 Three Mile Island and the 1986 Chernobyl accidents, the aging nuclear plants in countries with established nuclear-power production, and the emerging interest of other countries in acquiring their first nuclear plants, demands that urgent attention be paid to ensure that the likelihood of a major accident remains very low. The 2011 Fukushima Daiichi accident will also prompt needed attention to ensuring that nuclear facilities are adequately protected against natural disasters such as massive earthquakes and tsunamis. Nuclear safety and plant performance have benefited from the cultivation of a culture that makes safety the highest priority among plant personnel. Concern for security, however, has lived in the shadows for decades, mainly as a way to protect these potential vulnerabilities from adversaries. With threats of massive, destructive terrorism following the attacks of September 11, 2001, nuclear-plant owners, operators, and guard forces, as well as regulatory agencies, have taken steps to improve security. Like a major accident at a nuclear power facility, a terrorist attack would likely have widespread damaging effects on the global nuclear industry. Similarly, any misuse of commercial nuclear-power technologies, largely to make nuclear weapons, could lead to greater proliferation of nuclear-weapons programs throughout the world. A final challenge for the nuclear power industry is how to manage the disposal of nuclear waste so as to protect the public and the environment for thousands of years.

This book explains the benefits and risks of nuclear power in an accessible and authoritative manner. Beginning with an introduction to the fundamental science of nuclear energy, the book then covers the essential issues of enhancing energy security, financing of nuclear plants, combating climate change, preventing the proliferation of nuclear weapons, improving the safety of nuclear facilities from human-caused accidents or natural disasters, strengthening the security of nuclear facilities, managing radioactive waste, and creating a sustainable energy system.

ACKNOWLEDGMENTS

My journey to understanding the nature and possibilities of nuclear energy started around the time of the Three Mile Island accident. I was living not too far away, in Bedford County, Pennsylvania. Voraciously reading both science and science-fiction books as a teenager, I desired to become a scientist; but growing up in rural Pennsylvania, I had few role models. Fortunately, a couple of years after the accident I began to take classes from J. David Popp, who taught calculus, chemistry, physics, and nuclear science in my high school. For the latter course, the class had access to Penn State's research reactor. I am indebted to Dave for inspiring me to learn more about nuclear science.

As a nuclear engineering officer in the U.S. Navy, I experienced the standard of excellence required to make nuclear safety a top priority. Serving on a nuclear submarine at the end of the Cold War, I also came to understand the potential for the catastrophic demise of civilization by nuclear war. These two experiences shaped my career path and introduced me to an amazing network of people who were working to safely use nuclear power peacefully and who were striving to prevent nuclear war and further proliferation of weapons-usable nuclear technologies. Among the many people whom I have met and worked with, I would like to especially thank John Ahearne, Gene Aloise, Joshua Batkin, Michele Boyd, Jack Boureston, Matt Bowen, Frank "Skip" Bowman, Peter Bradford, Robert Budnitz, Alex Burkart, Ron Burrows, Richard Cleary, Thomas Cochran, Kevin Crowley, Marvin Fertel, Jacques Figuet, Richard Garwin, Lisa Gordon-Hagerty, Marc Humphrey, Shirley Jackson, Gregory Jaczko, Carol Kessler, Paul Lettow, Glen Levis, David Lochbaum, Micah Lowenthal, Edwin Lyman, Peter Lyons, Allison Macfarlane, Richard Meserve, Patte Metz, Rodney Nichols, Ivan Oelrich, Miles Pomper, William Potter, Vic Reis, Scott Sagan, Amy Sands, Frank Settle, Francis Slakey, Harold P. Smith, Henry Sokolski, Leonard S. Spector, Jack Spencer, Sharon Squassoni, Warren Stern, Michael Telson, Michal Vidard, Frank von Hippel, Fred Wehling, and Richard Wolfson. I am especially thankful to Carol Kessler and Warren Stern for hiring me to work in the U.S. State Department's Office of Nuclear Safety and for introducing me to international nuclear energy experts.

During the time I was at the Council on Foreign Relations (CFR), I was fortunate to have worked with two extraordinary research associates, Lisa Obrentz and Michelle M. Smith. Michelle co-wrote three articles with me on nuclear energy issues, as well as served as the writer and producer of CFR's multimedia Web-based nuclear energy guide. When I was completing this book at the Federation of American Scientists, I benefited from research assistance and discussions with Lindsey Marburger.

This book originated at the Council on Foreign Relations. I am very grateful for the leadership shown by CFR President Richard N. Haass and Director of Studies James M. Lindsay in forming a science and technology program and in providing encouragement for the nuclear energy project. Gary Samore, who had also served as the CFR's Director of Studies during part of this project, gave intellectual support and helpful advice while I was writing the Council Special Report, "Nuclear Energy: Balancing Benefits and Risks."

The project would not have occurred without the partnership of Washington and Lee University. Frank Settle, a dear friend and chemistry professor at Washington and Lee, was a joy to work with as co-principal investigator. I am also thankful for his bringing into the partnership the highly talented educational organization the National Energy Education Development (NEED) Project, led by Mary Spruill, and the work done by retired physics teacher Larry Skeens and NEED's Hallie Mills, in developing nuclear energy curricula for middle and high school teachers. This intellectual collaboration would have come to naught without the generous support of Harold "Gerry" Lenfest and the Lenfest Foundation. I am also deeply appreciative of the funding provided by the Richard Lounsbery Foundation.

I am grateful to David McBride for shepherding the book through Oxford University Press. He and assistant editor Alexandra Dauler asked insightful questions about and provided excellent editing of the manuscript. I am especially thankful for my agent Martha Kaplan for her pragmatic advice and help in keeping the project on track.

Most of all, I could not have completed this book without Catharine Ferguson, my wife who, many years ago, taught me the principles of clear writing (any muddled writing herein is solely my shortcoming). She gave me her love and support throughout the book project.

NUCLEAR ENERGY

WHAT EVERYONE NEEDS TO KNOW®

1

FUNDAMENTALS

What is energy, and what is power?

In the language of science, energy and power are distinct but linked concepts—that is, power is the rate of change of energy. But in common speech, people often think these terms mean the same thing. One way to think about energy is as the ability to do work. For example, work would involve moving an object such as a brick from a low to a high point. To move the brick, a bricklayer's body converts chemical energy obtained from food into the kinetic energy—energy of motion—of the body's muscles. These muscles then do work in lifting the brick. During this entire process, energy was neither created nor destroyed but only transformed from one form to another. So, another way to think about energy is as a fundamental substance of the universe. The other fundamental substance is matter—the material stuff that makes up the atoms of our bodies, the earth, the stars, and any physical object in the universe. As we will learn in this book, nuclear energy comes from transforming matter into energy. Metaphorically, we can think of matter as "frozen energy," which is melted and released by nuclear processes as energy that people can use to do work.

The total amount of energy and matter is conserved in the universe, meaning that the amount that exists at the beginning of an activity such as moving a brick is the same as the amount that exists after completion of the activity. But transformations among different forms of matter and energy occur continuously. Often in everyday speech we talk about "consuming" and "producing" energy. This is inaccurate, but as long as we are aware of what is really going on, we can continue to use these colloquial concepts.

Power relates to energy as the rate of changing energy from one form to another. A rate measures how fast or slow something changes over time. Thus, power equals energy divided by time. For example, a light bulb powered by 100 watts requires delivery of 100 joules of electrical energy per second in order to light the bulb. A joule is a common unit of energy used by physicists. As physicist Richard Wolfson has shown, a person of average strength can light a 100-watt bulb with relatively hard effort by turning a hand crank connected to an electrical generator. Averaging throughout the day and night, a U.S. citizen uses about 10,000 watts of power, including power used by agriculture, industry, residences, government, and transportation. Therefore, a typical American requires the equivalent of 100 "energy servants" turning these hand cranks. A typical European, in comparison, uses about half the amount of power.

While the watt is a useful measurement unit for light bulbs, most often kilowatts are used to measure the electrical power in the home, and megawatts are used to describe the capabilities of power plants. *Kilo* means "thousand," and *mega* refers to "million." A large nuclear reactor generates at least 1,000 megawatts of electrical power. This amount can be described as one gigawatt, where a *giga* means "one billion." Electric utility bills denote energy usage in *kWh*, which stands for "kilowatt-hour." This is an energy unit because it multiplies a unit of power (kilowatt), which is energy divided by time, by a unit of time (hour). To visualize the kilowatt-hour (kWh), it is helpful to remember that the chemical energy in one gallon of gasoline equals about 40 kWh.

This book will use the terms "nuclear energy" and "nuclear power" interchangeably when referring to the general topic of nuclear energy. Please keep in mind, though, that energy and power are interconnected but different concepts.

Is most energy that people use derived from nuclear energy?

Although it may seem like an outlandish claim, this is technically true. While this book will focus on the potential for people to use nuclear energy by commercially harnessing fission and fusion—two direct sources of nuclear energy—we should realize that the majority of the world's energy comes from solar energy, which originated as nuclear fusion energy.

Most energy sources in use today are fossil fuels: coal, oil, and natural gas. As the word *fossil* implies, these are ancient energy sources, formed eons ago. They have resulted from ancient life, such as prehistoric animals and plants decaying. Living things are largely made of the elements of hydrogen, carbon, and oxygen. Hydrogen and carbon can link through chemical bonds to form various hydrocarbon compounds. The simplest hydrocarbon compound is made from one carbon atom bonded to four hydrogen atoms. This compound is called methane, a primary constituent of natural gas. Longer chains of hydrogen and carbon make up petroleum and coal. After lots of heating and applied pressure from geological forces, the hydrogen and carbon in decaying, once-living matter is transformed into fossil fuels, which can be acquired through mining and drilling.

Still, what do fossil fuels have to do with nuclear energy? The ancient plants grew through photosynthesis, a process that takes carbon dioxide from the air and uses the energy source of sunlight. The sun produces light from nuclear energy. Deep inside the sun, very hot temperatures and very high pressures allow hydrogen to fuse together to form the next heavier element of helium (which, if the temperatures and pressures are intense enough, can also undergo fusion). Nuclear fusion releases energy. The highly energetic light from fusion is eventually emitted from the surface of the sun as visible light as well as other, nonvisible forms of light. A small fraction of this sunlight lands on the earth.

In addition to fossil fuels, which have trapped ancient sunlight, people can use today's solar energy for generating electricity, powering vehicles, and providing heat. For example, solar photovoltaic panels and concentrated solar energy to produce steam are two ways to harness the sun to generate electricity. Through photosynthesis, plants such as corn and sugarcane grow and produce sugars, which can be fermented to make biofuels to power cars and trucks. Solar water heaters can provide hot water for homes and business uses.

Wind also results from the sun's nuclear energy. As sunlight heats up the earth's surface, especially in the tropical zone, the air in these warmer regions becomes less dense than the cooler air in surrounding regions. The differences in air densities cause high and low pressure volumes of air. Wind is air moving from high to low pressures, and thustransporting excess heat from warm regions to cooler regions. People can harness the wind to generate electricity using turbines.

Solar and wind are considered renewable energy sources in that, as long as the sun shines, these sources will be available. And the sun is predicted to shine for about another 5 billion years. People have used only a tiny fraction of the available solar and wind energy. They have also hardly tapped geothermal energy, which is heat from the earth's interior. Geothermal energy results from radioactive decay of heavy elements such as uranium and thorium. Radioactive decay is a type of nuclear energy.

One may be wondering if there is any energy source that is not derived from nuclear energy. The answer is yes. Hydropower, for instance, is generated from flowing water. Gravity is the force behind this flow. Similarly, tidal forces, which can be harnessed to generate electricity, result from gravitational forces.

What is the origin of nuclear energy?

The sources of nuclear energy are fission, fusion, and radioactive decay. Nuclear fission occurs when certain types of heavy atoms become unstable and split into two medium mass parts, and nuclear fusion occurs when light atoms are forced together to make heavier atoms. Radioactive decay happens when unstable atoms emit energy in order to become more stable. All three processes involve interactions among powerful forces and changes of mass into energy.

The story of our understanding of nuclear energy began about 2,400 years ago in ancient Greece, when Democritus, a pre-Socratic Greek philosopher, conjectured that the world is made of indivisible substances he called "atoms." While modern science has shown that matter indeed is composed of atoms, scientists have learned that these atoms are divisible and can be split into separate parts. The two main parts are the nucleus, which is the atom's core, and the cloud of electrons that envelope the nucleus. The nucleus contains two types of particles: protons and neutrons. Protons are positively charged particles, and they exert forces on other charged particles like other protons or negatively charged electrons. Unlike-charged objects, such as an electron and a proton, attract each other, whereas like-charged objects, such as two protons or two electrons, repel each other. These are electrical attractive and repulsive forces.

If these were the only forces governing atoms, nuclei would burst apart because the protons inside a nucleus would push themselves away from each other. But this does not happen. So, there must be an attractive force that keeps a nucleus glued together. This is called the "strong nuclear force." Neutrons also feel the strong nuclear force and thus help contribute to keeping the nucleus together. But because they are uncharged, they do not feel the electrical force.

The key to understanding nuclear energy involves the push-and-pull between the repulsive electrical force and the attractive strong nuclear force inside the nucleus. If the repulsive force overpowers the attractive force, the nucleus becomes unstable. An unstable object can become more stable by changing its amount of energy or mass. For example, a ball on top of a hill is less stable than a ball at the bottom of a hill. The ball feels gravitational attraction from the very massive earth. Gravity tries to pull the ball down the hill. A slight nudge can cause the ball to roll down the hill. If the ball rolls down the hill and eventually settles at the bottom, it goes from a position of high-gravitational potential energy to a position of low-gravitational potential energy. This lower position is more stable than the higher position.

The stability of a nucleus is determined by how tightly bound together are the protons and neutrons inside it. Very stable nuclei have very tightly bound protons and neutrons. Nuclear scientists can quantify the amount of stability by measuring the binding energy of the nucleus. This is calculated by totaling all the masses of the nucleus's protons and neutrons, assuming that they were not bound inside the nucleus, and then subtracting the mass of the nucleus. An individual neutron or proton always has greater mass outside a nucleus than inside. The mass difference comes about because, when the neutrons and protons bind together, a tiny fraction of their masses is transformed into energy. Energy and mass are precisely related. According to Einstein's famous equation, energy equals mass times the speed of light squared. Because the speed of light is a big number, and because the equation says to multiply this number by itself, the tiny mass is multiplied by a huge number, thus resulting in a relatively significant amount of binding energy.

With this knowledge of binding energy comes an understanding of how nuclear energy is emitted. The next important concept is that the nuclei of various types of atoms have different binding energies. Iron, which is a medium-mass element, has the most tightly bound nuclei and thus the greatest binding energies. The nuclei of the lightest elements, such as hydrogen, helium, and, lithium, generally have the lowest binding energies. The nuclei of the heaviest elements, such as uranium and plutonium, have lower binding energies than the nuclei of iron but generally greater than the binding energies of the nuclei of the lightest elements. These observations led to the insight that there are two routes toward forming tightly bound nuclei.

The first route is to combine light nuclei together to form a heavier—more tightly bound—nucleus. The difference in binding energies between the heavy nucleus and light nuclei is the emitted nuclear energy resulting from the combination of the light nuclei. This combination occurs through a fusion reaction in which the attractive strong nuclear force overcomes the repulsive electrical force and thus pulls the two light nuclei together. The energy of nuclear fusion is distributed among the kinetic energy (energy of motion) of the heavy nuclei and the neutrons that are the output of the reaction, as well as gamma radiation, which is highly energetic light that is not visible. (In addition, neutrinos, which are uncharged and almost mass-less particles, are produced and carry away a small portion of the fusion energy, but because these particles are very weakly interacting with matter, their energy will not be available to be captured and used for producing work, such as electricity generation.) Later, you'll learn about the different methods of causing fusion and the difficulties of harnessing it to generate electricity.

The second route to releasing nuclear energy is to split less tightly bound heavy nuclei to form more tightly bound medium-mass nuclei. In particular, a heavy nucleus splits into two medium-mass nuclei and also outputs two to three neutrons. The medium-mass nuclei are called "fission products." The difference in binding energies between the medium-mass nuclei and the heavy nucleus is emitted as nuclear fission energy. *Fission* means splitting something into two or more parts. Although fission may occur spontaneously with a small probability in some types of heavy, unstable nuclei, typically the trigger for fission is the absorption of a neutron—similar to the nudge given the ball at the top of the hill. This absorption causes the nucleus to vibrate and then split. Similar to fusion reactions, fission reactions distribute the emitted energy among the kinetic energy of the fission products and neutrons, as well as gamma radiation. (Neutrinos also are produced and carry away a small portion of the fission energy, but as with fusion, this will not be useful in producing electricity.) Later, you will learn the different ways to cause fission reactions to occur in either nuclear bombs or nuclear reactors.

Nuclear physicists have discovered, through developing theories and performing experiments, that fission is most apt to occur in heavy nuclei that have an odd atomic mass number, which is the sum of the protons and neutrons in a nucleus. For example, the nucleus of a uranium-235 atom has 92 protons and 143 neutrons. Because each proton and neutron has one atomic mass, adding them up equals 235 atomic mass units, indicating why uranium-235 has that number label. Uranium-235 is known as "fissile material," owing to its relative ease to fission. Other fissile materials are americium-241, plutonium-239, and uranium-233. In contrast, even atomic mass nuclei such as thorium-232, plutonium-238, and uranium-238 are not fissile but are fertile, in that by absorbing a neutron they can transform into fissile material. This transformation occurs through radioactive decay.

Radioactive decay is the final process for releasing nuclear energy. Unlike fission and fusion, which depend on the strong nuclear force, radioactive decay is governed by the "weak nuclear force," which is called such because it is weaker than the strong force. Unstable nuclei can become more stable by decaying or transforming to more stable types of nuclei. Almost all fission products, for example, are unstable and will eventually decay. Depending on the type of fission product, this decay may occur very rapidly, in less than one second, or may happen in millions of years. During radioactive decay, energy is released as ionizing radiation, which is discussed in the next section.

What is "radioactivity"?

Energetically unstable isotopes release excess energy to become more stable. (Isotopes of an element have the same number of protons but differing numbers of neutrons in their nuclei. Each element has various numbers of isotopes, and each isotope is a unique combination of protons and neutrons.) This energy can be released, or "radiated," in different forms. An unstable isotope is known commonly as a radioisotope because it emits radiation and thus is "radioactive." For example, some radioisotopes emit a rapidly moving helium nucleus, which is made of two protons and two neutrons bound together. This helium nucleus is termed "alpha radiation."

In 1899, physicist Ernest Rutherford, working in Great Britain, discovered this type of radiation along with beta radiation while investigating the ability of uranium to ionize gases. (Uranium was used because previous experiments by French physicist Henri Becquerel showed that this element was emitting penetrating radiation. These experiments are described later in this chapter.) Ionization means that the radiation's energy can produce ions (charged atoms) by knocking negatively charged electrons off of neutral atoms. Rutherford gave alpha radiation its name after the first letter of the Greek alphabet and because it has an ionization capability more potent than beta radiation, named after the second letter of the Greek alphabet. A beta particle can be either an electron or a positron, the latter being a positively charged electron. Beta particles are typically emitted with a lot of energy, and so they travel at fast speeds when leaving a nucleus. While alpha radiation has a greater ionization potential than beta radiation because an alpha particle has twice the amount of electrical charge as a beta particle, beta radiation tends to be more penetrating than alpha radiation because of its higher speeds and its smaller charge. For example, a sheet of paper can block most alpha radiation, and a few sheets of aluminum, which is denser than paper, can block most beta radiation.

In 1900, a third type of ionizing radiation was discovered and was called gamma radiation, after the third letter of the Greek alphabet. Gamma radiation, unlike alpha and beta radiation, is uncharged. Specifically, gamma radiation is highly energetic light, which is much more energetic than visible light. Like all light, gamma radiation travels at the fastest speed of the universe, the speed of light, which is almost 300,000 kilometers per second (about 186,000 miles per second). Because gamma radiation is very fast and highly energetic, it is very penetrating and typically requires very dense materials such as a layer of lead or thick concrete to block. In addition to alpha, beta, and gamma radiation, an unstable nucleus can emit other types of radiation to become more stable. It can release protons and neutrons as well. All these emitted particles can ionize, or strip electrons from, atoms.

Why is ionizing radiation a health concern?

Too much ionization of bodily tissues could cause cancer and other harm to human health. Radiation-induced cancers typically take years to develop and become noticeable. The probability of developing cancer depends on the amount of ionizing radiation received. Thus, low doses of ionizing radiation have low probabilities of resulting in cancer. The good news is that the human body can often defend itself against the effects of relatively low doses of ionizing radiation. This means that very few people will actually develop cancer from exposure to low doses of ionizing radiation. But the unsettling news is that doctors and health physicists cannot be certain whether a person has developed a cancer as the result of exposure to low doses of ionizing radiation. The uncertainty is even more encompassing in that the health effects of low doses are controversial. That is, some researchers posit and cite evidence for beneficial health effects for very low doses. Other researchers instead claim that there is evidence for a threshold dose amount in which any doses less than the threshold would neither harm nor help health. Presently, public health policy is to minimize any excess exposure above natural background radiation. This policy is known as the "linear no-threshold" model, in which the health effects are assumed to be linearly proportional to the amount of ionizing radiation exposure.

Given a large enough sample of the population that has been exposed, scientists can predict what fraction of those people will develop cancer, although they cannot predict which individuals will develop cancer. To put this in perspective, people are exposed throughout their lives to many potentially cancer-causing substances, complicating the ability of health professionals to pinpoint an exact cause for a cancer unless there is clear evidence that a person has been exposed to relatively large amounts of a hazardous substance or has manifested a type of cancer associated with a specific cause.

High doses of ionizing radiation can cause more immediate health effects. These are called "deterministic effects" because they are directly determined by and proportional to the amount of high doses of ionizing radiation a body receives. One of the first noticeable deterministic effects is radiation sickness, with symptoms of nausea and vomiting. Higher doses can cause diarrhea and loss of hair. Even larger doses can severely damage the immune system and cause hemorrhaging. If the body cannot repair this damage, the person will die. Exposures to such high doses of ionizing radiation could happen to people close to nuclear explosions or powerful unshielded radiation sources, such as relatively large amounts of cobalt-60 or cesium-137.

Peaceful nuclear power, when operated safely, results in extremely tiny amounts of ionizing radiation being released into the environment. In fact, a typical operating coal-fired power plant emits more radioactive material in its exhaust than an operating nuclear power plant. Coal often contains uranium from the mined ore. Understanding the basic physics of isotopes, radioisotopes, radioactivity, and ionizing radiation is essential for making sense of the ongoing debates about disposal of nuclear waste and the potential hazards of an accident at or an attack on a nuclear power plant or other nuclear facility.

What is radioactive half-life, and how can knowing it help increase a country's security?

"Half-life" is a measurement of how long it takes for half a sample of a unique radioactive substance—that is, a radioisotope—to decay by emitting radiation. For example, tritium, a heavy form of hydrogen, has a half-life of 12.3 years and decays to helium-3 via beta radiation emission. The half-life is a property of the collective sample of a radioisotope. This means that one cannot predict when each individual nucleus of a radioisotope will decay; but given a large enough sample size, one can predict how long it will take for half of that sample to decay. Think about individual coin flips. One cannot predict, for a fair coin, whether a flip will result in heads or tails. But for a large number of flips, one can reliably predict that half the time there will be heads and half will be tails. In the case of radioactivity, things are more complicated in that there are thousands of different radioisotopes with many different modes of decay (ways to emit radiation) and many different half-lives. Each radioisotope has a definite half-life for each mode of decay.

The decay modes and the half-lives serve as fingerprints that allow investigators to specify exactly what radioisotope is being observed. This method is part of forensic analysis. It also allows radiation detectors at border crossings and other control points to pinpoint the particular radioisotopes that may be resident in a cargo container. Ceramic tiles, bananas, and kitty litter, for example, are some of the common substances that contain naturally occurring radioactive materials. Developing more sophisticated detectors to discriminate among these harmless materials and radioactive materials that make up nuclear bombs and radiological bombs (popularly known as "dirty bombs") can help protect countries from nuclear and radiological attacks.

How much more energetic is nuclear energy compared to chemical energy?

Burning fossil fuels releases chemical energy. Chemical bonds hold the atoms together in the molecules. When the hydrocarbon molecules that make up fossil fuels such as methane, petroleum, or coal are burned, the chemical bonds are broken between the hydrogen and carbon atoms. The chemical energy in these bonds is measured in units of electron volts. An electron volt is the amount of energy an electron acquires by being accelerated by a 1-volt battery. Because chemical bonds involve electrical forces among electrons and protons, the energy in a typical chemical bond is on the order of a few to several electron volts.

Fission and fusion reactions are far more energetic than chemical reactions because of the far superior strength of the strong nuclear force compared to the electrical force. On a per-mass basis (pound for pound or kilogram for kilogram), nuclear reactions release more than 1 million times more energy than chemical reactions. For example, the fission of one uranium-235 nucleus releases about 200 million electron volts. Even considering the fact that a uranium-235 atom is more than ten times the mass of a molecule of methane, you can see that a fission reaction is more than a million times more energetic than a chemical reaction.

Fusion is even more energetic than fission on a per-mass basis. Fusing deuterium and tritium—the two types of heavy hydrogen—releases more than 17 million electron volts per reaction. Because a deuterium nucleus is a proton and a neutron, and a tritium nucleus is a proton and two neutrons, their combined mass is only five atomic mass units. Thus, this fusion reaction releases more than 3 million electron volts per unit mass compared to a fission reaction's release of just less than 1 million electron volts per unit mass. Because of this fact and because deuterium is relatively abundant, nuclear scientists and engineers would make available a vast amount of energy if they could commercialize fusion energy.

Why has it been so difficult to commercialize nuclear fusion?

An oft-told joke in energy research is that nuclear fusion is the energy source of the future and always will be. While scientists and engineers have produced uncontrolled fusion reactions with nuclear weapons, they have struggled to harness this power source in a controlled manner. The main difficulty is in maintaining the required intense temperatures and pressures. There are three methods: gravitational confinement, magnetic confinement, and inertial confinement. The massive amounts of matter in the sun create a large enough gravitational field to confine the fusion reactants. But it is not possible to replicate the gravitational confinement of fusion on the earth. There is simply not enough matter.

Magnetic confinement involves using very powerful magnetic fields to trap the reactants. The fusion reaction can easily disturb the magnetic field, but this is still a method of active research. Several laboratories around the world have investigated magnetic confinement fusion. The project on the grandest scale using this technique is the International Thermonuclear Experimental Reactor (ITER), which is headquartered in Cadarache, France. The ITER involves China, France, the European Union, India, Japan, Russia, South Korea, and the United States. It is a multibillion-dollar, several-decades-long project. Construction work on some of the ITER's facilities took place in 2010, but the project organizers have been experiencing budget tightening, which may further delay the construction. The goal is to achieve first production of the plasma—the ionized gas needed for the fusion reactions—in November 2019.

Inertial confinement, the third method, directs an intense pulse of energy on a pellet of fusion fuel. This pulse should cause the pellet to implode with enough pressure and temperature to fuse together the deuterium and tritium. (The deuterium and tritium reaction has been the focus of fusion research because it is the relatively easiest reaction.) This method has shown some progress at the National Ignition

Facility (NIF) at the Lawrence Livermore National Laboratory in Livermore, California. The NIF apparatus fires 192 laser beams onto the fusion pellet. In January 2010, scientists at the NIF were able to heat a gold capsule to almost 6 million degrees Celsius. The next step will likely involve heating an actual pellet of fusion fuel. Even if that is successful, the challenge is to keep the fusion reaction going by feeding in and heating multiple pellets sequentially. The NIF's primary purpose is not to achieve commercial nuclear fusion but, instead, to provide a laboratory to simulate thermonuclear reactions inside advanced nuclear weapons.

The above methods are all "hot" fusion in that temperatures of millions of degrees are required to cause heavy hydrogen or other materials to fuse. By contrast, cold fusion, according to its proponents, creates fusion under room temperatures. On March 23, 1989, Martin Fleischmann, a world-renowned chemist, and Stanley Pons, a chemist, stunned the world when they announced that they had generated fusion in an experiment involving electrolysis of heavy water using a palladium electrode. "Electrolysis" refers to running an electric current through water to dissociate water into its hydrogen and oxygen components. Their evidence for fusion was an anomalous amount of heat produced that they claimed could only be caused by nuclear energy. Disillusionment soon set in when many researchers could not duplicate the results. Nonetheless, more than twenty years later, some scientists continue to study cold fusion, although most scientists remain highly skeptical.

How was nuclear fission discovered?

In 1932, British scientist James Chadwick discovered the neutron. This fundamental insight into the physical world was an essential discovery, but it was not enough to uncover fission, although scientists were quick to realize that bombarding substances with neutrons could create different substances. In particular, this conjecture led to the search for producing substances heavier than uranium, then the heaviest known element. In the mid-1930s, some nuclear scientists suspected that it may be possible to split, or fission, heavy elements. This speculation spurred competition among four major research teams. The team leaders were Ernest Rutherford in Britain, Frédéric and Irène Juliot-Curie in France, Enrico Fermi in Italy, and Otto Hahn and Lise Meitner in Germany.

The team in Germany was the first to demonstrate fission of uranium. But its victory was hardly assured because of the tribulations that Meitner experienced. She was born into a Jewish family in Vienna, Austria. Although as an adult she converted to Christianity, she was not immune to Nazi persecution of people with Jewish ancestry. Having overcome tremendous prejudice against female scientists, she eventually secured in the 1930s a position in Berlin at the prestigious Kaiser Wilhelm Institute for Chemistry. There she, as a physicist, collaborated with chemist Otto Hahn. Although several scientists with Jewish family backgrounds fled Nazi Germany as Adolf Hitler came to power in 1933, she stayed in Germany, protected by her Austrian citizenship. But when Hitler annexed Austria in 1938, Meitner fled in July that year to the Netherlands and then made her way to Sweden.

Remarkably, she was able to maintain scientific correspondence with Hahn, who was working with German chemist Fritz Strassmann on an experiment that bombarded uranium with neutrons. The two chemists discovered in December 1938 that barium, a medium-mass element, was an end product of the reaction. News of this discovery reached

Meitner and her nephew Otto Frisch, a physicist. Meitner and Frisch were the first to explain the theoretical mechanism that causes fission. They published this insight in January 1939. This publication opened the floodgates to research on fission. Meitner herself saw the implications for nuclear weaponry, but when asked to work on the Manhattan Project, she said, "I will have nothing to do with a bomb!" Meitner was denied the highest honor for her leading role in explaining fission. In 1944, Otto Hahn received the Nobel Prize in Chemistry for discovering fission, but Meitner did not share this award. In 1997, a study sponsored by *Physics Today* concluded that this was "a rare instance in which personal negative opinions apparently led to the exclusion of a deserving scientist" from receiving the Nobel Prize.

What role, if any, did Albert Einstein play in the discovery of nuclear energy?

Although many people may believe that Einstein discovered nuclear energy, he did not work directly in this area of research. A result from his special theory of relativity did, however, provide the fundamental underpinning for why prodigious amounts of energy can be released from the tiny-mass of a nucleus. Einstein's crucial insight is that energy and mass are equivalent. Mass, in a sense, is frozen energy. As mentioned earlier, this result is the famous equation $E = mc^2$. In words, this literally means that energy equals mass times the square of the speed of light. Because the speed of light is so large, squaring it—that is, multiplying it by itself—produces a huge number: 90 billion kilometers squared per second squared. Thus, multiplying this number by even a tiny mass will result in a relatively large amount of energy.

Although Einstein did not know how to unlock this latent energy, his theory pointed the way for other scientists to discover how to do that. Fission and fusion reactions are the two mechanisms that can release the tremendous amount of latent energy in matter.

Because of Einstein's fame, he did play a role in bringing the possibility of nuclear bombs to the attention of President Franklin Roosevelt. Having left Germany in 1932, Einstein settled in the United States and worked at the Institute for Advanced Study at Princeton, New Jersey. After the discovery of fission, émigré scientists Leo Szilard, Edward Teller, and Eugene Wigner persuaded Einstein to lend his name and fame to two letters (one in August 1939 and the other in March 1940) to inform Roosevelt about the implications of recent discoveries in fission and the possibility of Nazi Germany's building nuclear weapons. Einstein did not, however, work on the Manhattan Project, which developed the first nuclear bombs.

What is a fission chain reaction?

Imagine a line of dominoes set up so that if one domino is toppled, it will topple another, and so on until all the dominoes will fall down one at a time. This is analogous to a chain reaction in a nuclear reactor. This chain begins with a fissile nucleus such as uranium-235 or plutonium-239 absorbing a neutron, which causes the fissile nucleus to split. Then the fission of this nucleus releases two to three neutrons. The chain reaction in a nuclear reactor is designed so that only one of these neutrons, on average, would lead to the fission of another fissile nucleus. Using the analogy of dominoes, only one domino would topple at each step of the chain reaction.

In contrast, an explosive chain reaction is designed to have an exponentially accelerating number of fissions over time. Imagine dominoes set up in a triangular shape so that the toppling of the domino at a tip of the triangle results in toppling two or more other dominoes. Each of those falling dominoes will knock down two or more dominoes and so on in geometrically or exponentially increasing fashion. Analogously, in a nuclear explosion, the fission of a uranium-235 or plutonium-239 nucleus releases two or three neutrons and all or most of these neutrons will fission other fissile nuclei and so on in an exponentially increasing way. Inside a nuclear bomb, this chain reaction spreads extremely rapidly so that within microseconds, enough energy is emitted to destroy the heart of a city. In the explosion, the number of fissions is about 10 raised to the power of 24. To put this huge number in perspective, recall that 1 million is 10 raised to the power of 6; 1 billion is 10 raised to the power of 9; and 1 trillion is 10 raised to the power of 12. So, the number of fissions is 1 trillion times 1 trillion. Try to picture the number of grains of sand in the world. Mathematicians at the University of Hawaii estimated this number as 7.5 billion times 1 billion. Thus, the number of fissions inside a nuclear bomb far exceeds the number of grains of sand in the world.

What is uranium, where did it come from, and how was it discovered?

Uranium is a weakly radioactive, heavy, metallic element that is useful in a number of civilian and military applications, including reactor fuel, nuclear weapons, ballast for airplane tails, and armor-penetrating ammunition. Naturally occurring uranium in the earth's land and seas consists of three isotopes that are from most to least abundant: uranium-238, uranium-235, and uranium-234. The numbers refer to the number of neutrons and protons inside each isotope's nucleus. Because uranium always has 92 protons, uranium-238 contains 146 neutrons; uranium-235 has 143 neutrons; and uranium-234 has 142 neutrons. Uranium-238 makes up the vast majority of natural uranium atoms with a concentration of 99.28 percent; next comes uranium-235 with a concentration of 0.72 percent; and the tail end is uranium-234 with a concentration of 0.0054 percent. The desirable isotope is uranium-235 because it tends to fission easily as compared to the two other isotopes. Another desirable isotope is uranium-233 because it too easily fissions, but it does not exist naturally because of its relatively short half-life. Uranium-233 can be produced by breeding it from thorium-232.

The proportions of natural uranium isotopes have changed over time. To understand why this is so, it is necessary to take a brief look at the formation of the earth. About 4.5 billion years ago the earth was formed by coalescing from interstellar material circling around a newly born star that humans call the sun. The sun was formed when hydrogen, helium, and other materials in a huge ball of gas was compressed together by the force of gravitational attraction. Around the sun, a gigantic disc of material swirled. The earth and the other planets condensed from this material. The material came from a mixture of hydrogen and helium derived from the origin of the universe in the Big Bang and elements ejected from supernova, which were exploding gigantic stars. These supernovas made all the naturally occurring isotopes heavier than iron-56. As mentioned earlier, all the elements leading up to and including iron can be produced through fusion. Fusion occurs inside stars over a long period of time—up to billions of years.

Supernovas eject a mixture of dozens of different elements and hundreds of different isotopes into the interstellar medium of space.

Uranium is one of those elements formed via supernova. When the earth was formed, there were considerably more of the non–uranium-238 isotopes, but the concentration has shifted in favor of uranium-238 because of the much longer half-life of this isotope. U-238 has a half-life of 4.47 billion years. In comparison, uranium-235 has a half-life of 700 million years; uranium-234 has a half-life of 246,000 years; and uranium-233 has a half-life of 159,200 years.

While humans have used uranium oxide since at least 79 C.E. to add color to ceramic glazes, scientists did not identify uranium as a unique element until 1789, when German chemist Martin Klaproth discovered it in the mineral pitchblende. Uranium obtained its name from the planet Uranus, which had been recently discovered by the astronomer William Herschel. In 1841, Eugene-Melchior Peligot first isolated uranium as a metal. Discovery of its radioactive properties had to wait until 1896.

At that time, French physicist Henri Becquerel had been experimenting with uranium's phosphorescent property. He had been exposing uranium to light in order to see the phosphorescence, or "glow-in-the-dark," low-intensity light that is emitted a delayed time after the absorption of higher frequency light. Because clouds obscured the bright sunlight needed for the experiment, Becquerel placed in a closed drawer the uranium compound and the photographic plate that he was going to use in the experiment. Later he decided to resume the experiment, but before doing so, he developed the photographic film and discovered that it had already been exposed. Becquerel concluded that some energetic substance was emitted by the uranium to expose the film. This substance was high-energy radiation that originated from the nucleus of the uranium. Becquerel's colleagues Marie and Pierre Curie began a series of experiments with uranium ore. Using chemical techniques, they isolated and thus discovered two radioactive elements: radium and polonium. Radium took its name from radioactivity. The Curies named polonium after Marie's homeland of Poland. Radium served as a workhorse element for decades after this discovery, providing radiation for a variety of applications, such as self-luminous paint for watch dials. This practice was discontinued after many workers developed cancer from exposure to radium. Since the 1950s, radium has been gradually phased out of much of commerce because of the use of reactor-produced radioisotopes including cobalt-60 and cesium-137.

What is plutonium, how was it discovered, and how hazardous is it?

Like uranium, plutonium is a heavy element and can be used to fuel nuclear reactors and nuclear bombs. Plutonium contains 94 protons in its nucleus. The isotope of greatest interest is plutonium-239, which can easily fission and thus can fuel reactors or bombs. It has a half-life of about 24,000 years. Plutonium has fourteen other isotopes. Notable isotopes are plutonium-238 (half-life of 87.7 years), which can be used to power nuclear batteries; plutonium-240 (half-life of 6,560 years), which is a major constituent of spent nuclear fuel; plutonium-241 (half-life of 14.4 years), which is also present in spent fuel and is fissile; and plutonium-242 (half-life of 374,000 years), which is an additional isotope in spent nuclear fuel. Plutonium-238 has powered more than two dozen U.S. spacecraft. The *Voyager* spacecraft, for example, used Pu-238 to provide power to send back images of Jupiter, Saturn, and Uranus.

Plutonium was named after Pluto, the ninth planet (which was recently demoted to less than planetary status) and the Roman god of the underworld. As befitting the dark side, the discovery of plutonium was shrouded in secrecy. American nuclear chemist Glenn Seaborg and his colleagues Joseph Kennedy, Edwin McMillan, and Arthur Wahl produced this element in 1941, at the University of California. Because of plutonium's importance in fueling some of the first atomic bombs, the discovery was not publicly announced until 1948. Seaborg and McMillan shared the 1951 Nobel Prize in Chemistry in part for their work on plutonium, as well as other transuranium elements.

Plutonium generally has negative connotations among the public, not just because of its affiliation with nuclear weapons but also because of the health hazards associated with it. External exposure (outside the body) poses very little health risk because the predominant type of radiation emitted by plutonium is relatively easily stopped by the dead layer of outer skin. The potential health threats arise from internal exposure through inhalation or ingestion because plutonium mostly emits alpha radiation, which is short-ranged but highly ionizing. (Gamma radiation emitted by plutonium-239, for example, is relatively weakly energetic.) Considering the inhalation pathway, relatively large and small particles of plutonium would tend to not stay trapped in the lungs. Particle sizes around 1 micron or 1 millionth of a meter in diameter can become trapped and could, if the dose is powerful enough, lead to lung cancer. Most plutonium ingested through food or water contamination passes through the body. However, a small fraction can become dissolved in the bloodstream and then move throughout the body and eventually become resident in bones, liver, and other organs for many years to decades.

Why can't nuclear reactors explode like a nuclear bomb?

Reactors are designed to prevent an exponentially accelerating chain reaction from occurring. One reason that nuclear reactors cannot explode is that the concentration of fissile material is too low in most reactors to allow the formation of an explosive chain reaction. Another design feature that prevents a nuclear explosion from happening is that reactors contain materials that absorb excess neutrons. Nonetheless, reactivity excursions could happen in certain reactor designs (such as in the original Chernobyl reactor design) under certain adverse conditions. These excursions could spike reactivity quickly but would not constitute a nuclear explosion. The Chernobyl accident is discussed in chapter 5.

What is the nuclear fuel cycle?

Making nuclear fuel for reactors requires several activities, which constitute the nuclear fuel cycle. Figure 1.1 presents a flowchart of the cycle's major steps. The cycle begins with mining uranium. Uranium is contained in various deposits on land and in seawater. Several types of minerals, including pitchblende, uranite, carnotite, autunite, uranophane, and tobernite, contain uranium. Phosphate rock, lignite, and monazite sands also can contain uranium.

Milling is next needed to separate uranium ore from the tons of materials the ore is mixed with. Milling produces uranium ore concentrate, which is often called "yellowcake" because of its yellowish color. The yellowcake by itself is not

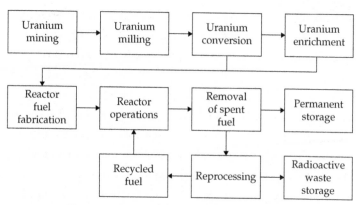

FIGURE 1.1: Flow chart of the nuclear fuel cycle.

usable in fueling reactors. So, it requires conversion to chemical forms suitable for producing natural uranium fuel or providing input material for uranium enrichment plants. These plants increase the concentration of the fissile isotope uranium-235 for making low-enriched uranium fuel. Later, the different types of reactors that require natural versus low-enriched fuel will be discussed, along with different enrichment methods.

Most reactors are fueled with the chemical compound uranium dioxide. To make the fuel, powdered uranium dioxide is pressed into fuel pellets. These are then stacked into long, thin rods. The rods are bundled together in a fuel assembly. To visualize a fuel assembly, picture a dozen or more long (up to several meters, or several yards, in length) and thin (up to ten millimeters, or about one-third of an inch, in diameter) fuel rods connected together by means of strong metallic brackets. The rods are spaced apart by several millimeters to allow coolant flow between them. Usually, a commercial reactor is fueled with several assemblies.

The fuel can stay inside the reactor for several months, up to a few years. During that time, uranium-235 undergoes fission and releases energy, which is used to generate electricity. The fission process also releases neutrons. Typically, two or three neutrons are released per fission. At least one of those neutrons is usually absorbed by a uranium-235 nucleus to continue the fission chain reaction. The other neutrons are absorbed by either the nuclei of uranium-238 in the fuel or the nuclei of other materials in the reactor. When a uranium-238 nucleus absorbs a neutron, it soon undergoes radioactive decay to produce neptunium-239, which later decays to plutonium-239, a fissile material. A significant portion of this plutonium fissions and thus contributes to energy released and electricity generated.

The fuel is removed from the reactor after significant amounts of the uranium-235 and plutonium-239 have undergone fission. This removed material is known as spent, irradiated, or used fuel. For consistency, we will call this material "spent fuel." After removal, the spent fuel requires cooling for usually a few years because it is very hot. Most of the heat is generated by the radioactive decay of fission products, the medium-mass isotopes produced by the fission of uranium and plutonium. The remainder of the heat comes from the radioactive decay of uranium and plutonium. Because the heat results from radioactive decay, it is known as decay heat. The spent fuel is placed in deep pools of water. The water acts as both a coolant to remove decay heat and a shield to protect workers from the intense radioactivity.

After sufficient reduction of the decay heat, the spent fuel may still be kept in the pool for several years more, or it may be placed inside dry storage casks, or it may be sent to a reprocessing plant. The decision depends on the policy of the government that has regulatory authority over the spent fuel. Here is where the fuel cycle branches into three different paths: the once-through use of uranium fuel, the single recycling of the spent fuel, and the multiple recycling of the spent fuel. (Later, in chapter 2, there's a discussion of why various governments have chosen different options.) In the first option, the spent fuel is destined for permanent storage at a disposal facility. But because no country has yet opened such a facility, spent fuel has remained in interim storage, usually at the power plants, in pools or storage casks.

Instead of disposing of spent fuel, governments could decide to reprocess it. Reprocessing is a set of techniques used to extract plutonium and unused uranium from spent fuel. The plutonium and unused uranium can then be recycled into new fuel. The highly radioactive fission products inside the spent fuel still require safe and secure disposal, however. So, reprocessing does not eliminate the need for waste storage and disposal. Rather than tens of thousands of years of storage for plutonium in spent fuel, most but not all of the radioactive fission products that result from fission of uranium and plutonium would decay within a few hundred years. In practice, however, the few countries, such as Britain, France, Japan, and Russia, that are reprocessing are typically doing a once-recycle, in that they extract plutonium from spent fuel that was initially only fueled with uranium; recycle the plutonium as fuel once; and store the spent fuel and other radioactive waste generated from that recycled fuel. Fully closing the fuel cycle, and thus consuming the long-lived radioactive waste, would require deploying a fleet of fast reactors that would burn up the plutonium and the other fissionable material. Multiple recycles would be needed to consume most of that material.

Unfortunately, reprocessing, as well as fast reactors, are not presently cost-competitive with the once-through uranium fuel cycle. (This issue will be discussed in more depth in chapter 2, on energy security and financing of nuclear power.)

Why are certain activities in the nuclear fuel cycle called "dual use"?

Because of the intertwined, or dual-use, nature of the nuclear cycle, Hannes Alfvén, a Swedish Nobel physics laureate, remarked, "Atoms for peace and atoms for war are Siamese twins." This means that the same technologies are useful for peaceful or military nuclear programs. The two technologies of concern regarding proliferation are enrichment and reprocessing.

Uranium-enrichment plants can produce either fuel for nuclear reactors or fissile material for nuclear bombs. In the case of enrichment plants making bomb-usable material, a plant would have to continue to enrich the concentration of uranium-235 to 90 percent or greater in concentration. In comparison, most low-enriched uranium fuel has a concentration of 3 to 5 percent uranium-235.

Reprocessing plants can also produce fissile material for recycled fuel or for nuclear bombs. But typically, the plutonium that is separated from commercial spent fuel is reactor-grade material, which is below weapons grade in its usefulness for nuclear bombs. Nonetheless, reactor-grade plutonium may still be usable in creating a nuclear explosion. Thus, reprocessing plants require adequate security and safeguards to prevent diversion of this plutonium to countries with weapons programs or to nuclear terrorists.

What are the various uranium-enrichment methods?

Several enrichment methods are available, but only one method, the centrifuge technique, has become the global standard. In the future, laser enrichment may offer commercial advantages if technical hurdles can be surmounted. Before discussing the centrifuge and laser enrichment methods, though, let's briefly look at the history of uranium enrichment to understand the likelihood of a country's using other, older methods to produce fuel for reactors or fissile material for nuclear bombs.

The first enriched uranium was produced during the Manhattan Project, which needed the highly enriched uranium for nuclear bombs. This bomb project used a combination of three methods: electromagnetic isotope separation (EMIS), thermal diffusion, and gaseous diffusion. In the first method, particle accelerators are used to separate uranium-235 from uranium-238 and thus increase the concentration of uranium-235. To start this enrichment process, uranium tetrachloride (UCl_4) is electrically heated to produce UCl_4 vapor. The vapor's molecules are then ionized. That is, a UCl_4 molecule is stripped of an electron to make it a positively charged ion. An electric field accelerates the ions to high speeds. A perpendicular magnetic field causes the accelerated ions to bend in a circular path. The lighter U-235 ions will follow a path with a shorter radius than the U-238 ions. The two types of ions then go through apertures leading to different collectors. Although this method seems easy to do, it is inefficient because usually less than half the feed material is converted to uranium ions and less than half of those ions are actually collected. Moreover, the process is time-consuming and requires hundreds to thousands of the EMIS accelerators and large amounts of energy. The United States used calutrons, a type of EMIS device, at the Y-12 plant in Oak Ridge, Tennessee, in the early 1940s. Although all five recognized nuclear-weapons states had tested or used EMIS to some extent, this method was thought to have been abandoned for more efficient methods, until it was revealed that Iraq had pursued it in the 1980s and early 1990s. Iraq was able to readily access unclassified detailed designs of this method.

The second original enrichment method was thermal diffusion. It makes use of the principle that lighter substances tend to rise faster than heavy substances when they are heated. This method was used for a limited time at Oak Ridge, Tennessee, during the Second World War to produce approximately 1 percent U-235, which was then used as input material for the electromagnetic isotope separation method described above. The thermal diffusion plant was dismantled when the gaseous diffusion plant began operation during the Manhattan Project.

Gaseous diffusion relies on the principle of molecular effusion to separate U-235 from U-238. "Effusion" is the term to describe how a lighter gas travels faster through tiny holes than a heavier gas. The holes are in the barriers of the diffusion stages of the enrichment plant. A diffusion stage contains a barrier between high- and low-pressure volumes and a compressor to drive the gaseous molecules from the high- to low-pressure volume. Each diffusion stage is connected to other diffusion stages by piping. In a uranium gaseous-diffusion plant, there are two types of gas: uranium-235 hexafluoride, and uranium-238 hexafluoride. The name "hexafluoride" indicates that there are six fluorine atoms. These are chemically combined with one uranium atom to form a uranium hexafluoride molecule. The difference in velocity between the two gases is small (about .4 percent). So, it takes many diffusion stages to achieve even low-enriched uranium. Thousands of stages are needed to make sufficient amounts of enriched uranium for either reactors or bombs. These plants are also energy hogs. For example, the Georges Besse I gaseous-diffusion plant at Tricastin, France, consumes most of the electrical power output produced by its three large nuclear reactors. While the United States and France still have commercial gaseous-diffusion plants as of early 2011, they are working toward replacing these plants with the much more energy-efficient centrifuge method. In comparison, the new Georges Besse II centrifuge plant at Tricastin will allow France to use almost all of the three aforementioned reactors to provide electricity to homes and businesses instead of to the enrichment plant.

The centrifuge method uses the force that occurs when something is spun in a circular motion. Imagine a merry-go-round in which children are being spun around. The heavier children feel a greater force than the lighter children. Analogously, the spinning centrifuge rotor affects the more massive uranium-238 hexafluoride gas molecules more than the lighter uranium-235 hexafluoride gas molecules. The former tend to congregate near the wall of the spinning rotor. A scoop near this wall directs the uranium-238 gas to the depleted uranium part of the centrifuge plant. "Depleted" here means that the uranium mix has less uranium-235 than the concentration of natural uranium. While some uranium-235 gas also goes into the depleted scoop, a greater proportion of uranium-238 is directed into that part of the plant. Another scoop located away from the wall directs a slightly enriched concentration of the uranium-235 molecules along with less concentrated uranium-238 gas into the enriched part of the plant.

Like an individual diffusion stage in a gaseous-diffusion plant, an individual centrifuge provides only a limited enrichment capability. Thus, to achieve the desired enrichment level, whether low-enriched or high-enriched, many centrifuges have to be connected together via piping. Typically, a few dozen, or up to a couple of hundred, centrifuges are grouped together in an arrangement called a "cascade." An enrichment plant may have up to tens of cascades in total to produce the desired enrichment level and amounts of uranium-235. The linking of the cascades and the arrangement of centrifuges in each cascade determine whether a centrifuge plant is optimized to produce low-enriched uranium, useful for most commercial reactor fuel, or high-enriched uranium, useful for bombs. This is relevant to current events because inspectors at the Iranian enrichment plants need to know how the plants are designed in order to assess how close Iran may be to making nuclear weapons, if it chooses to do so.

What are the nuclear-proliferation concerns for uranium enrichment?

The greatest proliferation concern presently centers on Iran, which received the beginnings of its enrichment program from a nuclear black market. A. Q. Khan, often called the "father of the Pakistani bomb," established a clandestine network that operated from the 1970s to at least early 2004, when Khan was forced to confess on Pakistani television that he headed this operation. (Elements of this network may still exist.) The Khan black market had connections to more than a dozen countries in Africa, Asia, and Europe and had supplied Iran, Libya, and North Korea with the knowledge and equipment for centrifuge enrichment. (Nuclear proliferation will be discussed in more depth in chapter 4.) The important point here is that the Khan network illustrates the dual-use nature of uranium enrichment.

A future nuclear black market could develop for an even more proliferation-prone technology: laser enrichment. It is considered "proliferation prone" because a laser enrichment plant takes up less physical space—perhaps as small as a warehouse—than a centrifuge plant, making the former harder to detect. There are two main methods of laser enrichment: atomic vapor laser isotope separation (AVLIS) and molecular laser isotope separation (MLIS). In AVLIS, a powerful, pulsed laser shines on a vapor of atomic uranium (that is, pure uranium that is not chemically combined with other elements). The laser light, if tuned to the proper wavelength, will selectively excite uranium-235 and ionize it. The ionized uranium-235 will be collected by a negatively charged plate. While the principle is relatively easy to understand, technical challenges have blocked commercialization of AVLIS.

The MLIS method involves a few different steps that do not require detailed explanations because they have the same core concept as AVLIS—that is, to separate uranium-235 from uranium-238 by laser excitation. The major difference between MLIS and AVLIS is that MLIS is used to excite uranium-235 hexafluoride and AVLIS is used to excite atomic uranium-235. As with AVLIS, MLIS involves technical difficulties that have stymied commercialization. However, as of early 2010, the Silex technique, which is an MLIS method, has shown some promise, according to the company developing it for commercial use. Silex was invented by Michael Goldsworthy, in Australia. The Australian and U.S. governments have a classified agreement to keep this method a secret. Global Laser Enrichment Corporation, based in Wilmington, North

Carolina, is attempting to commercialize Silex. Success may spur other nuclear companies to develop this or similar laser enrichment methods.

What is the thorium fuel cycle?

While uranium forms the basis of the commercial nuclear fuel cycle, thorium could eventually become the element central to an alternative fuel cycle. Swedish chemist Jons Jakob Berzelius discovered this element in 1828, and he named it after the Norse god of thunder, Thor. Thorium, a silvery white metal, is the ninetieth element on the periodic table. Thorium-232, the most common isotope, decays very slowly, with a half-life much longer than the age of the earth. Unlike fissile uranium-235, thorium-232 cannot readily fission; but like uranium-238, it is fertile and can produce fissile material. Thorium-232 can be transformed into fissile material by absorption of a neutron. When this absorption happens, thorium-232 becomes thorium-233, which is energetically unstable and decays relatively quickly, with a half-life of 22 minutes, to protactinium-233. This isotope then decays rapidly with a half-life of twenty-seven days to uranium-233, which is fissile and long-lived and can be used as reactor fuel. Uranium-233 is very efficient at fission because it produces more neutrons per fission on average than uranium-235 or plutonium-239. Thus, once a source of neutrons begins to transform thorium-232 into uranium-233, the fission of uranium-233 can provide sufficient neutrons to keep the reaction going. But because uranium-233 is such an efficient fissile material, it can be useful in powering nuclear bombs.

However, thorium reactors have been designed to make it hard to extract uranium-233 for bombs. For example, Alvin

Radkowsky, who was the chief scientist for the U.S. Navy's nuclear propulsion program, designed the Radkowsky Thorium Reactor, which does not require separation of uranium-233 from thorium-232 and protactinium-233, and thus has properties resistant to weapons proliferation. To increase proliferation-resistance, enough uranium-238 can be included in a reactor to dilute the concentration of fissile uranium-233. In addition, the generation of uranium-232 during the reaction increases proliferation resistance because this isotope is highly radioactive and poses a health hazard to anyone who would try to seize the uranium to make a nuclear bomb.

Another promising aspect of thorium is that it is more abundant in the earth's crust than uranium. So, in principle, it could supply hundreds of years' worth of electricity. Moreover, thorium-based reactors generally produce less long-lived radioactive waste than reactors fueled with uranium-235. With these significant benefits, one would think that scientists and engineers would have built and operated a large fleet of thorium reactors. But barriers have stymied the thorium fuel cycle. The presence of highly radioactive uranium-232 increases safety costs. Thorium-228, a potent alpha radiation emitter with a relatively short half-life, also complicates safe handling of irradiated fuel. Perhaps the biggest stumbling block has been that uranium-235 fuel had a substantial head start and the nuclear power infrastructure is set up for this fissile material. While uranium-235 remains relatively abundant, thorium will likely lag behind. India, though, has an incentive to commercialize thorium because it has abundant amounts of thorium but very limited indigenous supplies of uranium. But since India was given access to the international uranium market in 2008, the incentive to figure out how to commercialize the thorium fuel cycle has diminished.

How does a nuclear reactor generate electricity?

At its core, the nuclear reactor has a heat engine. Inside a heat engine, an energy source heats up a working fluid—typically water—to high temperature. This fluid is circulated in a loop, taking energy from the core via the heated fluid and returning to the core as cooled fluid. The heated fluid is either allowed to change phase from liquid to gas or is kept under high pressure to prevent the phase change. In the former case, the gas, typically steam, is directed onto a turbine, which is basically a large cylinder with fanlike blades attached to the outside of it. In the latter case, the superheated, high-pressure water transfers energy in a steam generator to another loop of water to make steam. The steam in this secondary loop is directed onto the turbine.

The hot gas impinges on the turbine's blades, causing them and the cylinder to turn rapidly. Attached to the turbine are tightly wound coils of wire, which is made of an electrical conductor. Powerful magnets are located near the wire. Turning wire in a magnetic field generates electrical current. This principle was discovered by British scientist Michael Faraday through experiments in the 1820s and early 1830s. The electrical current is then sent via the grid to homes and industries. So, a turbine and electrical generator convert mechanical energy to electrical energy.

Most of the energy from the reactor is not used to make electricity. For the present generation of commercial reactors, the efficiency of converting the nuclear energy to electrical energy is about 33 percent. That is, only one-third of the nuclear energy ends up as electricity. The rest of this energy is heat. Typically, this heat is sent into the environment and is thus wasted. Often, one hears the term "waste heat." Much of that heat could be used for other purposes, such as residential heating—what is called district heating—or industrial heating. But because nuclear reactors are usually located relatively far from urban areas for safety considerations, it is rare for reactors to provide residential heating. Industrial heating remains a largely untapped resource.

How many people's electricity demands can be supported with one large nuclear reactor?

A large reactor can provide at least 1,000 megawatts, or 1 billion watts, of electrical power. In the United States, the combined power rating of all of the 104 reactors is equivalent to about one hundred 1,000-megawatt reactors. That is, all of the country's reactors could generate up to 100,000 megawatts of electrical power. These reactors provide almost 20 percent of the electricity for a population of just over 300 million people. So, each reactor can generate the electricity demands of about 600,000 Americans. This number of people is equivalent to the population of a medium-size city. For example, just over 600,000 people live inside the borders of Washington, D.C.

What are the different types of nuclear reactors used for electricity generation?

A reactor designer has to make several choices: whether to slow down neutrons or use fast neutrons in the reaction; what type of coolant to use to keep the reactor core from melting down; and whether to pressurize the coolant to keep it liquid, let it boil in the reactor vessel, or use a gas as the working fluid in contact with the core.

The main reason to slow down neutrons is to enhance the likelihood of fission. Slow neutrons, or what are also termed "thermal neutrons," have a higher probability of causing fission of uranium-235, the main fissile source of reactor fuel, than fast neutrons. Slowing down neutrons requires a moderator. A moderator is made of substances that relatively quickly interact with fast neutrons and slows them to thermal energy levels. Imagine a billiard ball colliding with another billiard ball. The stationary ball is made to speed up, and the moving ball is slowed down. Imagine, too, that the balls are on a frictionless surface, so the moving ball can transfer energy only through collisions with other balls. It typically takes only a few collisions for the moving ball to transfer its energy and slow down appreciably.

When this analogy is applied to reactors, we see that a neutron can rapidly transfer its kinetic energy through collisions with another neutron or a proton, which has almost the same mass as a neutron. The transfer of kinetic energy happens most effectively between objects of the same mass because of the way the mathematics work out in the laws of conservation of momentum and conservation of energy. Because of this fact, substances with single protons in their nuclei produce an optimum slowing down of neutrons. One such common substance is water, which is made of two hydrogen atoms and one oxygen atom. Each hydrogen atom contains one proton in its nucleus. Water is abundant and can serve as a coolant as well as a moderator. Reactors that use a moderator are called "thermal reactors" because they rely mainly on fission from thermal energy neutrons.

The vast majority of the commercial reactors presently operating are thermal reactors. These reactors typically use water to moderate and cool the reactor core. But some designs use water mainly as a coolant and graphite—a form of carbon—as a moderator. For example, the so-called Chernobyl-type reactor, also known by the acronym RBMK, has this design characteristic. Eleven of these reactors are operating in Russia.

For those reactors using water as both moderator and coolant, the choice of type of water strongly affects the choice of fuel. Ordinary water is called "light water" to distinguish it from "heavy water," made of two heavy hydrogen, or deuterium, atoms bonded to one oxygen atom. At first glance, light water appears ideal for slowing down fast neutrons rapidly because the proton that makes up a light hydrogen atom is almost the same mass as the neutron, as discussed in the previous paragraph. In comparison, the deuterium atom has twice the mass of a neutron because it has one proton and one neutron in its nucleus. So, the transfer of kinetic energy during the collision is not ideal. But the complicating factor is that light hydrogen has a stickiness for neutrons. That is, there is a small chance that every collision between a neutron and proton leads to these two particles combining and forming a deuterium atom. The problem is that neutrons are then lost for fission. In contrast, heavy hydrogen has a much lower chance of capturing a neutron. Weighing the trade-off between rapidity of energy transfer and stickiness for neutrons, one finds that heavy water is a better moderator. The result is that a heavy-water reactor has more thermal neutrons available for fission. This means that natural uranium with a relatively low concentration of fissile uranium-235 can be used to fuel this type of reactor.

On the other hand, a light-water reactor requires enriched uranium that has an increased concentration of uranium-235 to compensate for the fewer available thermal neutrons. In effect, one needs to have more "targets," or uranium-235 nuclei, for the fewer number of "bullets," or neutrons, to hit and thus sustain the nuclear reaction. Light-water reactors are the predominant class of commercial reactors, with all U.S. reactors in this class and more than 80 percent of global reactors in this category as well.

Canadian nuclear engineers have developed heavy-water reactors called CANDU, for Canadian Deuterium Uranium. In addition to Canada, India and South Korea have CANDU-type reactors. While CANDUs are typically fueled with natural uranium—the type of uranium found naturally occurring on land and in the sea—they are fuel-flexible enough to consume slightly and low-enriched uranium, as well as recycled material from spent fuel generated by light-water reactors. CANDUs do not have to be shut down to be refueled. This offers an advantage for continual power production. But it provides a potential pathway for diversion of plutonium because outside observers would not be able to know when refueling is taking place just by monitoring, for example, satellite images of a CANDU plant. Light-water reactors, in comparison, have to shut down in order to refuel. Thus, satellite images can show that light-water reactors are refueling by monitoring for the absence of the water vapor that is emitted from cooling towers during the reactor's operation.

Reactors that use water (either light or heavy) as a coolant can be further characterized by whether the water is pressurized or not. A pressurized-water reactor (PWR) uses two loops of water. Figure 1.2 depicts the major parts of a PWR power plant. The primary loop circulates water through the reactor core and applies enough pressure to prevent this water from boiling while allowing it to be heated to higher temperatures. As mentioned earlier, much of the energy of the

Containment Structure

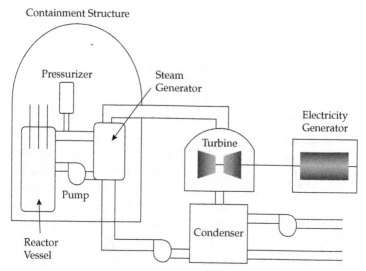

FIGURE 1.2: Schematic of a power plant including a pressurized-water reactor.

hot water in this loop is transferred in a steam generator to water in the secondary loop to make steam. After this steam turns the turbines, it is condensed back into liquid water and pumped by a feed pump to the steam generator, completing the cycle.

In contrast, a boiling-water reactor (BWR) uses one loop to transfer the reactor core's energy to the turbines. Figure 1.3 depicts the major parts of a BWR power plant. Above the core, but still inside the reactor vessel, the water is allowed to boil. As with the PWR, the steam from the BWR is condensed to liquid after it impinges on the turbines. In the United States, about two-thirds of the 104 commercial reactors are PWRs and the other one-third are BWRs. Worldwide, about 60 percent of reactors are PWRs, and a little more than 20 percent are BWRs.

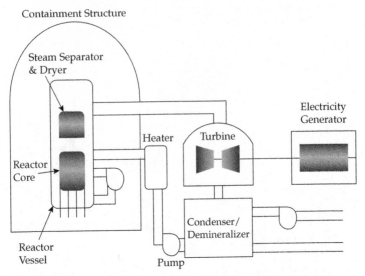

FIGURE 1.3: Schematic of a power plant including a boiling-water reactor.

Fast neutron reactors use high-energy neutrons to create fission. So, these reactors do not employ a moderator, but they do need a coolant to prevent the core from melting down and to transfer the heat from the nuclear energy to make electrical energy in a generator. Because fast neutrons are less than optimal for fissioning of uranium-235, fast reactors require highly enriched uranium, plutonium, or other types of fissile material that can make effective use of fast neutrons. Because a number of safety problems have plagued fast reactors, and because thermal reactors have been less expensive, very few fast reactors are currently operating. Proponents for completely closing the fuel cycle have called for fast reactors as necessary in order to burn up long-lived radioactive materials. While opinions differ on whether fast reactors will become safe enough and cost-effective, most experts agree that wide deployment of these systems are many years to decades away.

Finally, a few operating commercial reactors use gases as coolants. In particular, the British-designed advanced gas cooled (AGR) reactors employ carbon dioxide as the coolant and graphite as the moderator. In addition, some experimental work has been carried out on the Pebble Bed Reactor, an advanced gas-cooled, graphite-moderated reactor, which uses thousands of balls of uranium fuel embedded in graphite. This reactor was designed to be inherently safe, in that there is essentially very little risk of a reactor meltdown. Nonetheless, designers may still employ safety features used in traditional reactors in order to give the public more confidence in this design. Other next-generation reactors are described below.

Why were only a few types of reactor designs chosen for the present fleet of reactors?

Several different reactor designs have been conceived, but relatively few design concepts are widely employed. The choice of technology often depends on first movers and government support. In the early 1950s, when commercial nuclear power was just getting under way, the U.S. government had started investing in a submarine reactor program. Hyman Rickover, who was in charge of this program, needed compact reactors that could fit in the tight confines of a submarine. The nuclear navy soon settled on two designs: pressurized-water reactors and boiling-water reactors. And because the United States could make lots of highly enriched uranium for compact reactor cores, the navy decided on fueling its nuclear-powered ships with weapons-usable uranium. Commercial reactor designs emulated the basic principles of the navy's two designs, but instead they used low-enriched uranium fuel that was not weapons-usable. Perhaps if there had not been such a big push for nuclear-powered warships and the U.S. Navy had not chosen those two design concepts, the world's commercial nuclear fleet would have selected different designs that could have produced electricity more efficiently. In the future, the Generation IV reactors, as discussed later, may allow this and provide other advantages if they can overcome institutional inertia to switch away from well-established technologies.

What are the Generation IV reactors, and why are they considered potentially revolutionary?

The presently operating reactors are mostly considered Generation II, which are largely based on the first generation of reactors from the 1950s. Some Generation III reactors are starting to come online and are generally regarded as evolutionary improvements on Generation II. Truly revolutionary designs may be commercially available in the years approaching mid-twenty-first century. These designs are usually termed Generation IV. While some Generation IV designs build substantially on previous generations, the new technologies could offer significant breakthroughs, especially in safety and efficiency. In particular, the Generation IV Forum (GIF) has studied six major designs.

The supercritical water-cooled reactor would use water as coolant but at much greater temperature and pressure than a typical Generation II or III pressurized-water reactor. The higher temperature allows for much higher energy efficiency— about 45 percent as compared to 33 percent in a PWR. Because the water is in a supercritical phase and can be directed onto the turbine, there is no need for a steam generator and a secondary loop. This may result in big cost savings.

Second, the very high temperature gas reactor would use graphite for moderation of neutrons and helium gas as a coolant. One advantage over the current designs is that the helium is chemically nonreactive. And the very high temperature of about 950° Celsius, as compared to 315° Celsius for a Generation II reactor, means a more efficient reactor. The very high temperatures could provide large amounts of heat for industrial applications and for the generation of hydrogen. Copious amounts of hydrogen could power fuel cells in cars and trucks. Such vehicles would emit few or no noxious gases.

Third, gas-cooled fast reactors would employ high-energy or fast-moving neutrons to drive the chain reaction and would use helium gas for heat transfer from the reactor core. The temperatures would be high enough for both efficient electricity generation and hydrogen production. This type of reactor could be operated in a burner mode to consume long-lived fissionable materials or in a breeder mode to produce more plutonium for fuel. The former mode would hold out the potential to reduce the amount of radioactive waste needing to be stored. But the latter mode could, in principle, be misused to make fissile material for weapons.

Fourth, lead-cooled fast reactors would also use fast neutrons but instead would employ liquid lead or lead-bismuth to transfer heat from the core. These reactors could operate at either high or very high temperatures and thus could generate hydrogen. Another benefit of this design is its fuel flexibility. It can consume uranium, plutonium, or thorium-based fuels, as well as burn up other fissionable materials.

Designers have envisioned a wide range of power ratings, from smaller 300-megawatts of electrical power (MWe) units to large 1,400-MWe reactors. The former would offer the potential for connection to electrical grids in many developing countries.

Fifth, sodium-cooled fast reactors would transfer heat with liquid sodium. This design has already been used in a few countries and is not generally considered as revolutionary as the other Generation IV designs. The sodium cooling can pose a hazard in the event of a leak—sodium can easily catch fire. In 1995, at the Monju fast reactor in Japan, the secondary coolant system developed a pinhole leak. The resulting sodium reaction with oxygen in the air was contained but it forced the shutdown of the plant. Japanese nuclear industry authorities are still trying to obtain public confidence to restart the reactor. Its commercial restart has been delayed to at least 2014. In France, the Superphénix fast reactor was ordered shut down by Prime Minister Lionel Jospin in 1997, owing to "excessive costs" and poor operating performance. In addition, many of his political supporters were opposed to nuclear power in general and the Superphénix in particular. Russia is presently operating one sodium-cooled fast reactor. Also, India appears committed to moving ahead with this type of reactor, but the incentive to do so may change because of India's newly acquired access to the commercial uranium market for fueling thermal reactors.

Finally, molten-salt fast reactors are in the fast neutron family of designs but would employ liquid fluoride salts as coolant with the uranium fuel in the salt mixture. The reactor core temperature would be high enough for hydrogen production. Another variant of this design is to use graphite for some neutron moderation. The uranium fluoride salt fuel offers the advantage of producing no spent fuel assemblies. Moreover, the burn-up of long-lived nuclear waste may greatly reduce the high-level waste storage requirements.

Although all of these designs have undergone preliminary research, and in some cases have had operating experience, leaping to commercialization will likely require significant government investment. Industry will resist making this investment because of the huge upfront costs.

What can nuclear reactors do besides generate electricity?

Several navies have used reactors to generate electricity and provide propulsion for submarines and surface warships. A ship's nuclear reactor provides propulsion by producing steam to turn a turbine, which is connected via a shaft to the propeller. The Soviet Union, for example, had made more than 250 reactors for naval vessels. Most of these reactors were decommissioned after the end of the Cold War. Russia also has employed nuclear reactors to power civilian icebreakers in the Arctic region. Russian engineers have, in addition, created floating nuclear power plants that can be towed to coastal towns that have need of electricity.

Aside from seagoing reactors, scientists have used hundreds of research reactors to study the effect of neutrons and gamma radiation on materials such as satellites' components, to produce different substances through neutron-activation analysis, and to generate radioisotopes for research and commercial purposes. These reactors tend to have power ratings much smaller than commercial reactors that generate electricity. Research reactors with a rating of at least 25 megawatts of thermal power are of concern from the nuclear-proliferation standpoint. This size of reactor could produce enough plutonium for one nuclear weapon annually. Nuclear batteries made of certain types of radioisotopes can provide electrical power for remote locations such as lighthouses far from population centers and for space probes.

2

ENERGY SECURITY AND COSTS OF BUILDING POWER PLANTS

What is energy security?

The essential concepts of energy security are availability, reliability, and affordability. That is, more security flows from having many available sources of energy. Greater security also comes from having reliable sources, meaning that energy suppliers can be trusted to deliver those sources. Finally, energy security is linked to affordability in that suppliers will charge reasonable prices for energy supplies. Too low a price will provide inadequate revenue for the suppliers. Too high a price will tend to drive consumers to become more energy efficient and thus require fewer supplies. While the latter situation can favor the consumers, the suppliers can suffer unless they diversify their economies and thus not become overly reliant on revenue from supplying those energy sources. For example, leaders of the Organization of Petroleum Exporting Countries (OPEC) usually have striven to control production so that the price of oil does not stray too low or too high. Problems can arise when an OPEC member defects and increases production in order to win more profits at the expense of members who adhere to the production quota.

Energy-consuming states want to ensure that they are not overly dependent on one or even a few producing states. Leaders of these states would benefit by following Winston Churchill's advice that, "safety and certainty in oil lie in variety and variety alone." He made this observation when he was First Lord of the Admiralty just before the First World War, and he decided to switch the British navy from coal to oil. Oil offered advantages in its relatively high energy content per volume and its relative ease of refueling. The disadvantage was that, at that time, Britain did not have as abundant supplies of oil as it did coal. But Britain did not follow Churchill's dictum about variety and became more dependent on oil supplies from the Persian Gulf states, especially Iran. Britain then perceived that it had a strategic interest in controlling the Iranian government. The pursuit of this controlling influence culminated in August 1953, when Britain worked with the United States to overthrow Prime Minister Mohammad Mosaddegh in a coup and installed Shah Reza Pahlavi as a virtual dictator. The unintended consequence of this action has been bitter enmity to this day by the Iranian leadership and many Iranian people toward the U.S. and British governments. Thus, measures taken to secure energy supplies can have far-reaching geopolitical consequences.

In addition to seeking diverse external energy supplies, energy-consuming states can increase their energy security by developing the infrastructure to produce their own supplies. Brazil provides a recent outstanding example of a state that made the strategic decision to invest more in oil exploration, especially deep offshore exploration, and more in ethanol produced from sugarcane for transportation fuels. After a few decades of investment, Brazil has reaped the benefit of being essentially self-sufficient in providing for its transportation fuel.

Concerning nuclear energy, a state with nuclear power plants would want to have assured supplies of uranium and access to a diversity of fuel producers. Such a state would also seek a mix of suppliers for reactors. Even if the decision makers in that state choose only one or two reactor designs, they would first want to consider several designs, weighing the benefits and risks of each and using competition among reactor vendors to obtain a fair price.

Is energy independence feasible?

All countries are interconnected politically and economically. This interconnection extends to the realm of energy use and acquisition. While politicians can gain votes by making appeals to becoming energy independent, practically every country relies on other countries for energy sources or technologies to use those sources to make vehicular fuels, electrical generators, or residential and industrial heating and cooling systems. As long as markets for these goods and services function fairly, governments should have little concern about being denied energy supplies.

The biggest energy-supply concern today is the near-monopoly status that oil has on fueling cars and trucks. That is, drivers in most countries have very little choice in their fuel options other than gasoline or diesel derived from oil. Alternatives to fossil fuel include ethanol and biodiesel derived from plants. Although the United States, for example, has mandated a certain amount of production of ethanol as an alternative fuel to gasoline, the vast majority of cars sold in

American cannot now use more than 15 percent of ethanol in their fuel mixes. Ethanol is an alcohol that is mostly produced from corn in the United States, but it can also be produced from sugarcane and soybeans, and distilled from other sugar-laden crops and plant matter. A far-reaching technological breakthrough would be to make this fuel from nonfood plants such as switch grass, but it is more difficult to break down the cellulose in these plants.

Most of the fuel sold in the United States and many other countries is still gasoline. Only about $100 is required to convert most cars into flex-fuel vehicles that could consume a greater proportion of ethanol and methanol. While very few U.S. cars are flex-fuel capable, in contrast many Brazilian cars have this capability because of a Brazilian government decision. Another transportation fuel option that is emerging is electric-powered vehicles. These could be either pure electric powered or plug-in hybrids that have a backup gasoline-powered engine that kicks in once the electricity supply runs low. Giving consumers more choice in transportation fuel would not necessarily mean complete energy independence, but it would mean less dependence on oil-supplying countries.

Unlike the transportation sector, the electricity-generation sector offers consumers several fuel choices, including oil, coal, natural gas, nuclear, hydro, solar, and wind. No single source has a near monopoly influence. However, many countries rely on coal for a majority of their electricity, and an increasing number of countries are using more natural gas. Regarding reliance on nuclear energy, only a handful of countries generate 50 percent or more of their electricity from this source. France obtains almost 80 percent of its electricity from nuclear power; Belgium and Slovakia generate just over half of their electricity from nuclear power; Ukraine generates just under half of its electricity from nuclear. Most countries producing nuclear power need not fear overdependence on nuclear energy as long as they have other electricity-generation options and do not generate most of their electricity from this source. As discussed later, France and a few other countries address possible overdependence by investing in plutonium-recycling plants.

Have countries ever been shut out of the nuclear-fuel market?

Yes. Although the nuclear-fuel market has generally been reliable, a few countries have experienced fuel disruptions, have perceived that the market is unreliable, or have expressed concern that restrictions will be imposed in their ability to access nuclear fuel. Until the early 1970s, the United States was the single provider of commercial uranium-enrichment services for states outside of the Soviet Union, and only a handful of states controlled the major uranium mines. This situation spurred competition in the enrichment market so that today there are a few other major enrichment providers, including France, Russia, and the Urenco consortium of Germany, Great Britain, and the Netherlands. China is following suit. During the apartheid regime, South Africa was subject to multilateral sanctions and was denied nuclear fuel. As discussed more fully in chapter 4, on proliferation, until late 2008, India had been shut out of this fuel market because of sanctions resulting from its violation of safeguards on a research reactor that made plutonium for a 1974 detonation of a nuclear explosive.

Iran is the most visible example of a country being denied access to the nuclear-fuel market. In 1974, Iran under the rule of Shah Reza Pahlavi lent the French government $1 billion to help build an enrichment plant in France, and in 1977, Iran paid another $180 million to buy into ownership of Eurodif. Soon after the 1979 Islamic Revolution, the Iranian government decided not to pursue development of nuclear power and sued France to regain its money. Once the lawsuit was settled in 1991, giving Iran $1.6 billion for its investment plus interest, Iran still remained an indirect shareholder in Eurodif via a French-Iranian consortium named Sofidif. Iran at that time asked for delivery of enriched uranium because it had revived interest in nuclear power. France denied the request, pointing out that the contract had expired and that the 1991 lawsuit gave Iran no claim to enriched uranium from Eurodif. Iran has cited this experience as showing that international ownership of enrichment facilities does not function as proponents claim. On the other hand, France has reason, based on Iran's actions in this case, to believe that Iran may not be a reliable partner in financing future internationally owned enrichment plants.

Are European countries too dependent on Russian energy supplies?

Russia supplies a substantial amount of Europe's demand for natural gas. This demand is growing in more and more countries. The states that buy more than 50 percent of their natural gas from Russia include Austria, Bulgaria, the Czech Republic, Greece, Hungary, Slovakia, Slovenia, Turkey, and Ukraine. A number of these states—such as the Czech Republic, Hungary, Slovakia, and Ukraine—also depend on Russia for nuclear-fuel supplies.

Many analysts have argued that Russia depends on Europe as a consumer because energy sales are Russia's main source of revenue. Despite this dependence, Moscow has already deployed energy cutoffs. On January 1, 2006, Russia's Gazprom shut off natural gas to Ukraine because Russia claimed that Ukraine was not paying for all of its gas and was also diverting gas destined for use in the European Union. Although Ukraine initially denied this accusation, the Ukrainian national gas company later confirmed that it had diverted some gas for domestic use. Although Gazprom resumed gas supplies three days later, more than a dozen European countries were seriously affected. About 80 percent of the European Union's gas flows through Ukraine. In January 2009, another dispute between Russia and Ukraine led to another gas cutoff, which caused eighteen European countries to report gas shortages. European governments' concern about Russian gas reliability has driven support for projects to build alternative gas pipelines, find alternative sources of natural gas, and develop natural-gas strategic reserves as buffers against supply disruptions.

Increased nuclear energy use could also help these governments make themselves more resilient against disruptions of natural gas. Natural gas, however, is not only used for electricity generation but is also used for business and residential heating, fertilizer production, and cooking. Nonetheless, because electricity generation is increasingly using natural gas, alternative electricity sources could lower the demand for natural gas. But there are obstacles to increasing the use of nuclear energy in Europe, ranging from requirements to shut down Soviet-designed nuclear plants to political opposition to nuclear power. Bulgaria, Lithuania, and Slovakia have had to shut down some Soviet-designed plants as a condition to enter the European Union. The German coalition government in the late 1990s agreed to phase out Germany's nuclear power plants and not build new ones because of political opposition to nuclear power. Germany has also committed to taking a leading role in curbing greenhouse gas emissions. It relies heavily on coal-fired plants for electrical power. Germany plans to increase substantially its use of renewable energy sources such as solar and wind. Because building enough wind farms and solar power plants to replace the decommissioned nuclear plants will take many years and will be expensive, Germany will likely be forced to import more natural gas to meet its goals for reducing greenhouse gas emissions and to compensate for the phase-out of nuclear power plants. Germany, thus, risks reducing its energy security because of its greater reliance on imported natural gas, especially supplies from Russia.

What role has nuclear energy played in reducing certain countries' dependence on fossil fuels?

Nuclear power has reduced dependency on oil for electricity generation in the United States and France. This reduction of oil dependency started in the 1970s, a time of major change in electricity generation for a few countries. In the early 1970s, oil reached peak production in the United States. From then on, the United States' ability to pump oil from domestic sources would fall behind growing U.S. demand. Compounding the oil challenge, OPEC embargoed oil in response to the U.S. decision to supply Israel's military during the October 1973 Yom Kippur War. The embargo lasted until March 1974. Because of this oil disruption, the United States took steps toward requiring the auto industry to make more efficient cars and reducing the use of oil in home heating and electricity generation.

In the late 1970s, the United States had plans to build dozens of additional nuclear reactors. But because of increased energy efficiency, as described above, the economic downturn, and increased costs for nuclear plants, many of these reactors were canceled. Nonetheless, nuclear power played a positive role in reducing U.S. dependency on oil for electricity generation. In 1975, the proportional sources of U.S. electricity generation were as follows: coal at 44.5 percent, natural gas at 15.6 percent, hydroelectric at 15.6 percent, oil at 15.1 percent, nuclear at 9 percent, and nonhydro renewable at .2 percent. In 2004, in comparison, electricity generation was: coal at 51.5 percent, nuclear at 20.8 percent, natural gas at 16.3 percent, hydroelectric at 7.0 percent, oil at 3 percent, and nonhydro renewable at 1.5 percent, according to the U.S. Energy Information Administration. The proportional use of nuclear energy more than doubled in that thirty-year period while oil use shrank by a factor of five. This occurred despite the fact that the United States has yet to complete an order for a new nuclear plant in more than thirty years. Many reactors, however, that had been ordered in the late 1960s and early 1970s were built in the 1980s. The last completed reactor was Watts Bar I, in Tennessee, in 1996. In addition, U.S. nuclear power plants have increased their capacity factors. Capacity factor is the percentage of the year a reactor is operated at full power. At the time of the 1979 Three Mile Island accident, the plants had average capacity factors under 60 percent. This accident sounded the alarm about safety and performance. Today, because of these improvements, almost all U.S. nuclear plants have capacity factors greater than 90 percent. The third contributor to increased use of nuclear power was the raising of the power rating, which is the maximum power-generation capability, of many plants. Plant owners have invested in improved turbines and electricity generators, for example, to increase the power rating of their plants.

During the 1970s, France also faced a major decision about its future electricity generation. The slogan "no oil, no coal, no gas, no choice" captured the essence of the French energy conundrum. France has limited fossil-fuel resources and limited supplies of uranium. However, because uranium supplies were available in former French colonial countries in Africa and on the world market with far fewer restrictions than oil, the French government decided to undertake a huge nuclear-reactor building program. Throughout the 1970s and 1980s, France increased from a handful of reactors to dozens. Presently, it has fifty-eight commercial reactors and it is building a large 1,600-megawatt reactor at Flamanville. France generates more than three-fourths of its electricity from nuclear power, about 10 percent from hydroelectric power, about 12 percent from coal and natural gas, and a small fraction from nonhydro renewable such as wind and solar. Thus, France does not rely on oil for electricity, but like the United States, it consumes large amounts of oil-derived fuels for transportation.

How could nuclear energy further reduce dependence on fossil fuels?

Nuclear energy can make further reductions in fossil fuel use in two main ways: (1) displace coal, oil, and natural gas-fired electricity-generation plants; and (2) power transportation to replace gasoline and diesel derived from petroleum. In the first area, nuclear energy presently provides about 15 percent of global electricity. In comparison, fossil fuels provide 41 percent from coal, 20 percent from natural gas, and almost

6 percent from oil, for a total of about 67 percent. Consequently, nuclear energy has a huge potential for displacement of fossil fuels because nuclear power presently provides a relatively small proportion of the world's electricity in contrast to fossil fuels, and because nuclear fuel is relatively abundant. The challenge is to build reactors as fast as or faster than the demand for fossil fuel plants. In the second area, nuclear energy–generated electricity can charge up electric-powered cars and trucks, as well as public transportation such as subways and electric-powered buses and trolleys. But very few electric-powered vehicles are in use. This may change dramatically, but that will require many years to a few decades to make serious in-roads because of the relatively slow turnover in the use of vehicles and the time required for the auto industry to gear up to produce millions of these new vehicles annually. The other fuel option that nuclear power could help provide is hydrogen for fuel cells. While hydrogen is not an energy source, it is an energy carrier that requires significant amounts of energy to liberate it from water or other hydrogen-bearing substances. Nuclear power plants that operate at very high temperatures could provide the necessary energy to generate hydrogen. The hydrogen would then power fuel cells in electrical generators in cars, trucks, and other applications such as home or business electrical usage.

What countries use commercial nuclear power, and how much electricity do they obtain from it?

Thirty countries or territories use nuclear energy to generate electricity. (Taiwan is included as a territory because most other states do not recognize it as an independent nation-state.)

See table 2.1 for a listing of the states that use commercial nuclear power. (This information dates from February 2010 and relies on data compiled by the International Atomic Energy Agency and the World Nuclear Association.) The distribution of commercial nuclear power around the globe is uneven. Europe has sixteen countries—the most in the world—with nuclear power plants. France is the world leader in proportional use. In North America, Canada, Mexico, and the United States use nuclear energy. The United States has the most commercial reactors in the world with 104. Asia has the two most populous countries: China and India. Although both use nuclear power, their proportional use is small, but ambitious government plans by Beijing and New Delhi will likely stimulate much greater generation of nuclear energy.

Australia, Africa, and South America stand out as having little or no commercial nuclear power. Australia—the country that is also a continent—has huge supplies of uranium but no commercial reactors, although there has recently been some government and public discussion of developing them. In Africa, only South Africa has a nuclear power plant, but a number of other states including Algeria, Egypt, Libya, and Nigeria have expressed interest in acquiring these facilities. In South America, Argentina and Brazil possess nuclear power plants; Chile and Venezuela have stated some interest.

How many more countries are likely to acquire commercial nuclear power plants?

In recent years, dozens of countries have expressed interest in acquiring their first nuclear power plants. Some of these countries, such as Egypt and Turkey, have tried to do so in the past while others, such as Saudi Arabia and the United

TABLE 2.1: *Countries with nuclear power plants.*

State or territory	Nuclear-generated electrical energy, in 2009 (billion kWh)	Percentage of domestic electricity	Operable reactors in July 2010	Power-generation capacity in July 2010 (MWe)
Argentina	7.6	7.0	2	935
Armenia	2.3	45.0	1	376
Belgium	45.0	51.7	7	5,943
Brazil	12.2	3.0	2	1,901
Bulgaria	14.2	35.9	2	1,906
Canada	85.3	14.8	18	12,679
China	65.7	1.9	11	8,587
Czech Republic	25.7	33.8	6	3,686
Finland	22.6	32.9	4	2,721
France	391.7	75.2	58	63,236
Germany	127.7	26.1	17	20,339
Hungary	14.3	43.0	4	1,880
India	14.8	2.2	19	4,183
Japan	263.1	28.9	55	47,348
Korea (South)	141.1	34.8	20	17,716
Lithuania	10.0	76.2	0	0
Mexico	10.1	4.8	2	1,310
Netherlands	4.0	3.7	1	485
Pakistan	2.6	2.7	2	400
Romania	10.8	20.6	2	1,310
Russia	152.8	17.8	32	23,084
Slovakia	13.1	53.5	4	1,760
Slovenia	5.5	37.9	1	696
South Africa	11.6	4.8	2	1,842
Spain	50.6	17.5	8	7,448
Sweden	50.0	34.7	10	9,399
Switzerland	26.3	39.5	5	3,252

continued

TABLE 2.1: *(continued)*

State or territory	Nuclear-generated electrical energy, in 2009 (billion kWh)	Percentage of domestic electricity	Operable reactors in July 2010	Power-generation capacity in July 2010 (MWe)
Taiwan	39.9	20.7	6	4,927
Ukraine	77.9	48.6	15	13,168
United Kingdom	62.9	17.9	19	11,035
United States	798.7	20.2	104	101,263
Total	2,560	14	439	374,815

Arab Emirates (UAE), are relatively new to entertaining this notion. The UAE is one of the most likely newcomers to achieve this feat because it has substantial amounts of money to pay for reactors. In fact, in December 2009, it reached a deal with South Korea to purchase reactors. It will likely take up to ten years for the first of the reactors to begin operating because of the time required to establish a new regulatory authority; train operators, safety inspectors, and other personnel; and build a reactor. But one reason South Korea was chosen was its track record in building reactors relatively quickly. The most recent reactor in South Korea was built in just over four years. The other main reason was that South Korea offered the lowest price as compared to other vendors. Officials from Areva, the French nuclear company that was the main competitor, complained that the UAE made a mistake in opting for low cost over the safest design, which they claimed was Areva's EPR-1600. The added safety features in part increased Areva's costs.

Aside from the Emirates, it is uncertain how many other new entrants there will be. One of the most interesting developments is the clustering of these countries. Many Middle Eastern and North African countries with large Arab populations have expressed interest. Nonproliferation experts have observed that these countries' announcements of interest in nuclear power plants are strongly correlated with the growth of Iran's nuclear program. Even if the Arab states are not now explicitly trying to obtain nuclear weapons, they may try to leave the option open as a future deterrent against nuclear attack if Iran decides to acquire nuclear weapons. While it may strike some people as odd that these states are trying to obtain nuclear power when the region has copious oil and natural gas, it is important to recognize that the distribution of these resources is uneven. Jordan and Yemen, for example, have very little oil and natural gas. Although Saudi Arabia and the UAE have lots of these resources, their leaders say they want to free up more for export and to do so they need alternative energy sources such as nuclear.

Another clustering is in Southeast Asia, where Indonesia, the Philippines, Thailand, and Vietnam stand out as states that want nuclear power. In South America, Chile, Ecuador, and Venezuela have expressed interest. In Sub-Saharan Africa, Ghana, Namibia, and Nigeria may eventually join the club. But it is important to underscore that all the newcomers confront significant hurdles. They have to develop effective regulatory agencies; train legions of qualified personnel to build and operate the plants; instill a safety and security culture among regulators and the nuclear workforce; find substantial financial resources to pay for the plants, fuel, and maintenance; and commit to a decades-long investment. And active management of the radioactive waste from a plant may require more than a century.

How do the costs of nuclear plants compare to other types of power plants?

Because of choices in electricity generation, the cost of a nuclear power plant makes sense only in comparison to the costs of other electricity sources. The total cost of any source depends on capital needed for construction (including financing charges), operating and maintenance costs (including the price of fuel), and decommissioning and disposal fees. In addition, because of the security and safety concerns about nuclear plants, owners have to pay for guards, gates, and protective barriers to protect the plant against attack, as well as liability insurance to help cover the costs of an accident. Not all of these costs fall to the plants' owners. Governments help protect against airplane crashes on nuclear plants by investing in security at airports, and government police and national guards can serve as backup protective forces. Moreover, governments have capped liability coverage on nuclear accidents so that insurance fees for the power plants' owners are kept at relatively low levels.

Because nuclear power plants are designed to operate at or near full power for many months, they are considered base-load power sources. "Base-load" refers to the amount of electrical power needed to meet the level of demand experienced twenty-four hours per day, seven days per week. Electrical demands above this base-load are serviced by peaking power sources that can be turned on relatively quickly as demand fluctuates. For example, a plant run on natural gas can provide peaking power because it can respond rapidly. It can also supply base-load power. Coal-fired power plants are the other major base-load electricity source. Hydroelectric and geothermal electric plants also can provide base-load power, but a hydroelectric or geothermal plant can be built only in locations where there are adequate supplies of water or adequate sources of available heat from the earth.

A comparison of plant construction costs to the fuel costs for the three base-load sources of coal, natural gas, and nuclear shows that nuclear plants typically have the highest capital costs and relatively low and predictable fuel costs. Natural gas plants have the lowest capital costs but usually higher and more variable fuel costs. Coal plants tend to be in the middle in terms of both costs. Because relatively few nuclear plants have been built globally and especially in the United States in the past two decades, there is little recent construction experience to estimate reliably the capital cost needed to build new plants. A major point of contention is how to include the financing charges in the advertised cost of a plant. That is, should the full financing charges be added to the total sticker price of the plant, or should the so-called overnight costs—assuming fictitiously that the plant can literally be built overnight—be the figure of merit? The former is the more honest figure, but because it can be a very high price, often the latter value is quoted. With the understanding that there are many estimates of these prices, it is still useful to know the ballpark estimate. For the overnight cost, the 2009 Update to the Massachusetts Institute of Technology's Nuclear Power Study estimated $4,000 per kilowatt (based on 2007 U.S. dollars). Thus, a 1,000-MWe plant would have an overnight cost of $4 billion. The financing charges could add another $2 billion or more depending on the amount of time needed to build the plant and the interest rate charged. Because the newer reactors are larger than 1,000 MWe—upwards of 1,400 to 1,600 MWe—the total cost for a new large reactor could soar to around $9 billion. In comparison, the overnight costs for natural gas and coal plants are $850 and $2,300 per kilowatt, respectively. Because these plants can usually be built faster than nuclear plants, their overall construction costs are significantly lower. Even when the higher fuel costs of natural gas and coal are factored in, a nuclear plant's costs are greater. Recent increases in the availability of natural gas in the United States, thereby driving prices lower, have made natural gas plants even more cost-competitive compared to nuclear plants.

While the upfront costs of nuclear power plants are expensive, the plants are cheap to operate if the capital, fuel, and maintenance costs are amortized over forty to sixty years of operational life. Once the capital costs are paid off, nuclear plants are cost-competitive with coal and natural gas plants, the two other major base-load power generators. Federal loan guarantees can help reduce the upfront financial risks as long as taxpayers have adequate protections in the event of defaults on the plants' construction.

How can nuclear power plants be made more cost-competitive?

For better or worse, producers of every energy source have received subsidies and financial incentives, such as federal tax credits for the generation of renewable energy and tax deductions for the costs of oil exploration and building refining facilities. In a perfect world, these price distortions would be made as transparent as possible. This is not to argue that subsidies are always bad. They can do good, especially when an energy source that serves a public good, like reducing carbon emissions, is not cost-competitive with high carbon-emission sources.

The financial challenges of building and operating a nuclear power plant depend on where it is located. Governments such as China and the United Arab Emirates, which have substantial cash reserves, can relatively easily pay for nuclear plants. Moreover, governments such as China, France, and Russia, which have extensive control over national electricity generation, can often make decisions on plant purchases that affect their entire countries. In comparison, if the electrical utility in a less centralized system does not have large capital reserves of at least several tens of billions of dollars, it will not have adequate collateral to offer the investors in a nuclear plant. In that case, the utility can ask for loan guarantees from the government in order to reduce the financial risk to investors, but this may increase the risk to taxpayers if the likelihood of default on the loan is high. The U.S. government has in early 2010 offered about $8 billion of loan guarantees to the Vogtle power plant in Georgia, conditioned on the company's meeting licensing milestones. Tens of billions of dollars of loan guarantees are also available for other plants. While such guarantees have been gaining political support, many economists and other analysts have argued against these guarantees, claiming that they distort the marketplace and put too many taxpayer dollars at risk.

To further reduce financial risks, a utility's executives may seek to merge their company with other utilities, to create a large enough corporation with adequate market capitalization. Exelon, the U.S. utility with the largest share of nuclear plants, tried to merge with the utility NRG Energy, but was rebuffed. Exelon, however, does serve as an important example of how utility mergers can result in acquisition of additional nuclear power plants. In 2000, Exelon was formed from the merger of PECO Energy Company, based in Philadelphia, and Unicom, based in Chicago. Using the financial leverage of the merger, Exelon by October 2009 had full or majority ownership of seventeen nuclear reactors.

Alternatively, a utility may turn to foreign governments for support. For example, Constellation Energy in Maryland had received financing from Electricité de France (EDF) for a nuclear plant at the existing Calvert Cliffs nuclear power site, which is the closest commercial nuclear plant to Washington, D.C. Constellation Energy and EDF had formed the commercial consortium Unistar. In October 2010, however, Constellation Energy sold its shares in Unistar to EDF after it declined receipt of U.S. government loan guarantees. Although EDF officials have expressed interest in going forward with the construction of the plant, they may need to find another U.S.-based partner company in order to meet the legal requirements of the U.S. Atomic Energy Act, which prohibits foreign ownership of U.S. plants. Thus, foreign government support for U.S. nuclear power plants can face significant hurdles.

Another financial mechanism may be available for utilities in a regulated system. In such a system, the state or regional regulatory authorities may set a price on the generated electricity to make nuclear more competitive or may allow the utility to recoup many of the construction costs from utility customers before the plant even begins running.

In addition to government financing, federal loan guarantees, mergers, and regulatory incentives, the two other fundamental approaches are to make fossil-fuel alternatives more expensive and to lower the costs of nuclear plants. The two methods for discouraging use of fossil fuels are to levy a tax or fee on carbon emissions or to institute a cap-and-trade system. While economists tend to favor a carbon tax because it sends a clear market signal, politicians tend to prefer cap-and-trade because instead of a direct tax it sets a cap or limit on the total amount of carbon emissions and then allows emitters to trade emission permits. A major criticism of cap-and-trade is that its implementation may have many loopholes. On the positive side, through U.S. government enactment of the 1990 Clean Air Act, cap-and-trade has worked to cut the emissions of gases that cause acid rain. But this system was done on a much more limited regional scale—concentrating on the eastern half of the United States, rather than meeting the global challenge of greenhouse gas emissions. Although the word *tax* can upset people, the government may reduce the burden of a carbon tax by rebating almost all of it through people's income taxes. The government could accomplish this by either lowering income taxes by almost the same amount that it raised carbon taxes or by collecting all of the carbon taxes and then giving taxpayers a tax refund of almost the same amount collected. The government may set aside a small fraction, but still a large amount, of money from the collected fee in order to fund research and development of energy efficiency and low-carbon energy systems. To make the cost of low-carbon sources more competitive, governments have offered tax credits to producers of these sources. For example, the U.S. Energy Policy Act of 2005 offers 1.8 cents per kilowatt-hour for up to 6,000 megawatts of new nuclear capacity for the first eight years of operation. This credit equates to $125 million annually per 1,000 megawatts, or a total eight-year credit of up to $6 billion for 6,000 megawatts. Similar tax credits are available for wind and solar energy.

The other fundamental approach to making a nuclear power plant more cost-competitive is to reduce its construction costs. One method is to focus on only a couple of reactor designs and to replicate them. The first-of-a-kind design typically has a lot of unforeseen hurdles to overcome, thus driving up costs. For example, the first ever Evolutionary (or European) Pressurized Reactor being built by Areva in

Olkiluoto, Finland, has experienced a cost excess of more than 2 billion euros over its original budget of 3 billion euros and a schedule delay of at least three years. But with further construction of a particular design, the costs should start to come down with greater experience. Similarly, engineers can develop ways to speed up construction through using faster drying concrete and procedures to have more work done at the same time rather than sequentially. South Korea, for example, has implemented those methods. Another method is to not rely exclusively on large reactors. A large power reactor may cost several billion dollars. Instead, building small modular reactors may require smaller market capitalizations, making them more affordable. The cost disadvantage is that, on a per kilowatt basis, the larger plant tends to be less expensive. But when a utility executive has to decide on whether the company can afford the total price tag, the smaller reactor may be appealing. In addition, to meet increasing demands for electricity, small modular reactors may offer more flexibility with respect to scaling up.

Why is it difficult for the supply chain to keep up with forecasts of demand for new nuclear plants?

One of the most daunting challenges in any industry is to prevent the "bullwhip effect." This occurs when mismatches between inventory and orders get amplified along the supply chain, similar to the buildup of the wave along a bullwhip that has been cracked. With renewed interest in nuclear power plants around the world, supply companies are anticipating the demand for expanded capacity. The big question is whether the demand will stay constant, will increase, or even will decrease in the coming years. Considerable sums of money are riding on the demand forecast. For example, only one company, Japan Steel Works, has been making ultra-heavy forgings for large reactor pressure vessels. While this company's executives are ramping up its forging capacity, they are acutely aware that their company suffered economic losses for three years after 1998, when Germany decided to phase out its nuclear plants. Presently, Japan Steel Works has a long queue of orders. Additionally, the increased demand has spurred competition. Sheffield Forgemasters in Britain, for instance, has been retooling its facilities to make ultra-heavy forgings. To ensure that critical parts, such as pressure vessels and steam generators, are available, utility executives place orders years in advance of when the parts will be used. If their plans for new plants fall apart, the executives are counting on being able to sell their place in the queue to other buyers. Although this practice may recoup costs for the buyers, it may also send a false signal about the real demand for nuclear plants.

Industry officials are aware of potential supply-chain problems and have taken actions to help tame the bullwhip. Nonetheless, as research at MIT's Sloan School of Management has shown, rational people still tend to make errors in their judgment of when to order parts, what to keep in inventories, and how to anticipate dramatic changes in demand. One proven solution is to have extensive information sharing up and down the supply chain, especially to more effectively communicate demand. The increasing globalization of the nuclear industry may make matching supply and demand easier because of the lower barriers to transmitting proprietary information inside a conglomerate. During the past decade, American-based Westinghouse merged with Japanese-based Toshiba to form Toshiba-Westinghouse, and Hitachi in Japan bought out

U.S.-headquartered General Electric's nuclear division to form GE-Hitachi. In a vertically integrated business model, one company has access to all parts of the nuclear fuel cycle. Areva, the French nuclear industry giant, owns capabilities in uranium mining, uranium enrichment, fuel manufacturing, reactor construction, radioactive waste management, and plutonium recycling. Such a model allows, in principle, more responsive changes to demands for supplies. Similar to Areva, Russia's Rosatom State Nuclear Energy Corporation demonstrates vertical and government ownership. Areva has also bought stakes in other nuclear companies such as Mitsubishi in order to create more market dominance. If global demand keeps growing, Areva, Mitsubishi, Rosatom, GE-Hitachi, and Toshiba-Westinghouse, the traditional nuclear powerhouses, will likely face increasing competition from South Korea's Kepco, as well as from China and India.

How many skilled people are required to build and operate nuclear plants?

One of the concerns about expanding nuclear power too fast is the shortage of highly skilled people needed to build and run the plants. To get a handle on the numbers of people and types of jobs, the U.S. Department of Energy (DOE) in 2005 estimated the workforce for building eight reactors, with work starting in 2010 and finishing in 2017, and assuming that each project proceeded on a five-year construction schedule. This time frame has turned out to be too ambitious because the first new U.S. reactor would likely not even receive its license to begin construction until late 2011. Nonetheless, DOE's report usefully notes that the labor force during the most active period of construction of the reactors would top at about 8,000. During that time period, about four reactors would be under construction, meaning that approximately 2,000 workers are needed per reactor. Despite common assumptions that the labor would mostly consist of nuclear engineers and radiation safety physicists, these specialties constitute only a small portion of the workers. Most jobs would entail craft laborers such as welders, cement makers, and electricians. The DOE report specified the requirement of about 1,000 operation and maintenance staff, 200 quality-control inspectors, 400 construction inspectors, 500 construction engineers, 100 Nuclear Regulatory Commission inspectors, and 300 people to start up the plant. Because of the decades-long stagnation in reactor construction in the United States, practically all of the nuclear workforce positions are facing shortages. This lack is especially acute among nondestructive testing professionals, reactor operators, nuclear engineers, and radiation safety (health) physicists. These highly skilled professions require many years of training. The industry is also facing a wave of retirements of older workers who had entered the field decades ago, during the major boom period of building. So, turning out the workforce for a potential nuclear energy revival will not happen quickly. The nuclear industry also confronts competition from other energy industries. To meet this challenge, the U.S. government, other governments, and nuclear companies are investing more in recruitment and training.

Can construction of nuclear power plants keep pace with the increasing demands for electricity?

Presently, nuclear power generates about 15 percent of the world's electricity. Global electricity demand is likely to nearly double between 2010 and 2030. (This assumes business-as-usual improvements in energy efficiency. More efficient energy usage will lower demand for new power plants.) So, within the next two decades, usage of nuclear power would have to double to keep pace. This projected rate of growth means that every sixteen days, a new 1,000 MWe reactor would have to connect to the electrical grid. Although this is ambitious, it is not impossible; the world witnessed a similar rate of nuclear building during the 1980s—the heyday of construction. But to do so will require ensuring adequate capacity for providing personnel and parts for the plants, as discussed earlier.

Will the world run out of uranium and, if so, when?

The earth has an abundance of uranium on land and in the seas. The real issue involves how much it costs to gain access to the uranium deposits. Imagine owning a mining company. Mining is only profitable if you can earn more money than you spend on extracting the ore. The cost depends on the location of the ore, the concentration of uranium in the ore, the extraction method, and the demand for uranium. The ideal ores would contain high concentrations of uranium, require little chemical processing to extract the uranium, and reside near the earth's surface. A desirable concentration is 1 percent or more uranium in the amount of material extracted. Typically, the concentration is much less. To put the amount of ore in perspective, consider that a 1,000-MWe reactor requires about 25 metric tons of enriched uranium annually. Acquiring this material means mining and milling approximately 50,000 metric tons of ore to extract 200 metric tons of uranium oxide concentrate, which is enriched to the 25 metric tons of material for fuel.

While no one has done a complete survey of the world's available uranium, the Nuclear Energy Agency regularly publishes estimates of recoverable reserves depending on the typical price for uranium. Based on a price of $130 per kilogram (about $60 per pound) of uranium and the current demand for uranium at about 68,000 metric tons per year, the known recoverable resources are more than 4,700,000 metric tons. Consequently, the world would have enough uranium at these prices for the next seventy years, assuming present demand. Although this period of time may seem short, if the demand for nuclear power significantly increases, incentives for increased uranium mining and prospecting would likewise increase, leading to increased supplies. These are just the land resources. Experts estimate that the oceans contain enough uranium for several hundred years of nuclear power. However, sea extraction is currently much more expensive than land mining. Technological break-throughs may lower this cost, particularly if demand for uranium increases.

From the energy-security standpoint, the location of the uranium deposits fundamentally matters. If most of the world's uranium were controlled by tyrannical leaders, there would be cause for concern. Fortunately, deposits are spread out among many countries. And from the perspective of the United States, United Kingdom, and other major democratic, nuclear-power producing countries, allied countries contain substantial amounts of uranium. In particular, Australia and Canada are two of the top uranium suppliers. Of the countries with newly declared interest in acquiring nuclear power plants, Jordan stands out as having recently discovered large quantities of uranium. While China ranks among the major uranium-supplying states, its government has been making deals with other supplying states such as Kazakhstan, which is one of the top suppliers, to ensure security of supply because of China's plans for a vast expansion of nuclear power.

Finally, extended burn-up fuels, which are under continual development, can also stretch uranium supplies. That is, these types of fuels would make more efficient use of the uranium-235 as compared to traditional fuels.

Why have some countries pursued reprocessing of spent nuclear fuel for commercial purposes?

As discussed in the previous chapter, reprocessing uses chemical techniques to extract plutonium and other fissionable materials from spent nuclear fuel. Most countries with nuclear power plants do not have commercial reprocessing facilities, but France, India, Japan, Russia, and the United Kingdom have them. The decisions to pursue these facilities arose from the time when these countries believed they were facing dire uranium shortages. Because less than 1 percent of natural uranium is fissile, technologies such as reprocessing to allow use of plutonium for fuel offered the option of extending the supplies of uranium. Up until the mid-1970s, global supplies of uranium were estimated to be scarce. This perceived scarcity drove certain countries to pursue reprocessing. India, France, and Japan, for instance, have small amounts of indigenous uranium. However, all three states—now that the U.S.-India nuclear deal has been approved—have assured access to the international uranium market. They could also easily stockpile uranium, which is readily storable. Thus, on grounds of security of supply, none of the states that presently reprocess have a compelling reason to continue. The United

Kingdom, for instance, will likely leave the reprocessing business owing to its lack of profitability.

Some states appear committed to operating their reprocessing facilities because of the view that plutonium is or will become a valuable resource or because reprocessing may eventually provide for optimal nuclear-waste disposition. Regarding the first viewpoint, the price tag for making plutonium-based fuels exceeds the cost of uranium-based fuels. According to a 2003 study from Harvard University's Belfer Center, the cost of uranium would have to soar to $360 per kilogram ($164 per pound), which is almost three times the typical price, in order to make plutonium fuel from a reprocessing plant economically competitive with uranium fuel from an enrichment plant. Because the cost of nuclear fuel is a small portion of the total cost of nuclear power, French citizens, for example, pay about 6 percent more for their electricity because of the reprocessing costs. Based on a cost comparison, reprocessing does not win out over the once-through uranium fuel cycle. But economic inertia has kept reprocessing alive because of the massive sunk costs in building reprocessing plants. France and Japan, for example, have spent several billion dollars per plant.

Reprocessing proponents argue, however, that more is at stake than this cost comparison. They point out correctly that reprocessing reduces the volume of high-level radioactive waste that requires long-term storage. But reprocessing also makes lots of low-level radioactive waste. Moreover, in order to derive full benefit from the high-level waste reduction, reprocessing would have to be continued on the spent fuel resulting from the plutonium-based fuels. But countries that reprocess do not usually carry out reprocessing on that type of fuel. Instead, they store that spent fuel, in effect not benefiting from a significant reduction in the volume of high-level waste. This spent fuel is stored until the time when it may become economically and technically viable to build and operate many fast neutron reactors. These reactors can burn up fissionable materials such as plutonium, curium, and americium. Thus, in principle, fast reactors can help alleviate the problem of nuclear waste disposal by consuming many of the heavy, long-lived fissionable materials. These long-lived materials affect repository requirements because of the tens of thousands of years needed to contain them and because several fission products are soluble in water. Nonetheless, a well-sited repository should be able to meet such requirements.

France and Japan illustrate the difficulties of deploying fast reactors. As of 2010, France does not have even one fast reactor after recently shutting down the Phénix prototype reactor, and it has no plans to acquire another for several years. Similarly, Japan has struggled with fast-reactor technology. The Japanese Monju fast reactor experienced a sodium fire in the secondary part of the plant in 1995 and has since not received permission to operate commercially. To fully consume the fissionable materials, a fleet of fast reactors would be required with about one fast reactor for every two thermal reactors. For the world's current amount of about 440 thermal reactors, more than 200 fast reactors would be needed.

Meanwhile, the rate of consumption of plutonium in fuel for thermal reactors considerably lags behind the rate of separation of plutonium from spent fuel. In recent years, five to ten tons of excess plutonium have been accumulating annually. Presently, about 250 metric tons of plutonium have been separated from spent fuel—enough fissile material for thousands of weapons.

Why did the United States decide to not pursue reprocessing, and will it revive this practice?

In 1976, Republican President Gerald Ford took the advice of experts to issue a halt to the U.S. reprocessing program. He lost the presidential election that year to Jimmy Carter. In 1977, Carter reiterated this policy and decided that his administration would try to encourage other countries to refrain from reprocessing. The arguments for this policy were twofold. First, reprocessing is more expensive than the once-through uranium fuel cycle. In March 2010, an official with Areva estimated that a commercial-scale reprocessing plant in the United States would cost about $25 billion. Second, reprocessing is a nuclear weapons–usable technology and may result in further nuclear proliferation if it spreads to more countries. PUREX, the only reprocessing technique that is commercially used, is especially prone to proliferation. It completely separates plutonium from a protection barrier of highly radioactive fission products. Compared to these fission products, plutonium is weakly radioactive and could be handled as long as someone does not ingest or inhale appreciable amounts. So, a terrorist or a thief could, in principle, steal this material without killing himself in the process, assuming that there was not adequate security guarding the plutonium. Moreover, reprocessing in the United States would signal to other countries that it is okay to do this latent proliferation activity.

To take leadership in convincing other countries not to reprocess, the United States has refrained from reprocessing for more than thirty years. It has actively exerted political pressure on countries that do not already have reprocessing plants to not acquire them. In 2001, however, the U.S. government under President George W. Bush signaled that it was open to reprocessing as long as it could be done in a "proliferation-resistant" manner. Vice President Richard Cheney, in particular, led a U.S. government energy-policy study in early 2001 that advocated this approach. Nonproliferation experts expressed concern that even proliferation-resistant reprocessing would still pose too high of a risk. It may not pass the "Iran test." That is, the United States would not willingly give a proliferation-resistant reprocessing plant to a state that may have interest in making nuclear weapons. Such a state may be able to make a clandestine PUREX-type reprocessing plant that could use diverted material from a proliferation-resistant reprocessing plant. Alternatively, the state may alter the latter plant to allow production of weapons plutonium. Compounding this problem is that the proliferation-resistant methods being researched did not offer nearly the level of inherent protection against theft or diversion as plutonium surrounded by significant amounts of fission products, according to studies done by U.S. national labs.

Nonetheless, the Bush administration in 2006 launched the Global Nuclear Energy Partnership (GNEP), which sought to promote further use of nuclear power in a proliferation-resistant way. One aspect of GNEP was to research and develop proliferation-resistant methods of using plutonium. But because of the concerns about the continuing proliferation risk, GNEP aimed to restrict such work to the existing nuclear-weapon states and Japan, which already was reprocessing. Many proposed client-states in the developing world were opposed to this plan because it appeared to deny them their rights to the complete nuclear fuel cycle. GNEP policy architects then had to modify the proposal so as to underscore that no country's rights were taken away. Still, proponents of

GNEP did not want to see the further spread of reprocessing to countries that did not already perform it.

Reprocessing remains a point of active political debate in the United States. During the 2008 presidential campaign, for instance, Republican nominee John McCain accused Democratic nominee Barack Obama of not supporting reprocessing. McCain's view was that reprocessing offered the United States the opportunity to mitigate the nuclear-waste problem and to open up new fuel supplies. But Obama, who supports more use of nuclear power, did not see the need to rush toward deployment of reprocessing facilities. Instead, the Obama administration has supported research into proliferation-resistant reprocessing and has not foreseen the need for building reprocessing plants for decades to come.

3

CLIMATE CHANGE

What is the greenhouse effect?

Greenhouses are designed to keep plants warm even when outside temperatures are cold. A greenhouse does this by allowing visible light from the sun to enter through the greenhouse's glass, thereby capturing its energy. Material inside the greenhouse absorbs visible light and reradiates part of the light's energy as heat in the form of infrared light, which has a longer wavelength than visible light. When this longer wavelength light encounters the molecules of the glass, they absorb the light and radiate some of it out of the greenhouse while the rest of it is radiated back inside the greenhouse. The result is that the interior becomes warmer.

The earth's atmosphere itself is like a greenhouse. Visible light from the sun penetrates the atmosphere. Materials that make up our planet absorb this light and reradiate it partly as infrared light. Certain molecules in the atmosphere tend to readily absorb infrared light and reradiate it, thus trapping part of this heat inside the atmosphere. The atmospheric molecules that work well as greenhouse gases are carbon dioxide, water vapor, and methane. These gases occur naturally. For example, animals exhale carbon dioxide as a process of metabolism. Plants absorb, or "inhale," carbon dioxide as part of their living process. The slight excess of carbon dioxide from this natural system has allowed the earth to have a greenhouse effect conducive to life. Without the natural greenhouse effect, the average surface temperature would be about -19° Celsius (-2° Fahrenheit)—too cold to support life. Presently, the atmosphere has a carbon dioxide concentration of only about .04 percent. The global average surface temperature is around 14.6° Celsius (58.3° Fahrenheit). More greenhouse gases dumped into the atmosphere tend to increase the warming effect. Essentially, any gas molecule that has three or more atoms acts as a greenhouse gas. Molecules in this size range absorb infrared light and block part of its transmission into outer space.

Human activities have increased the concentrations of carbon dioxide, as well as other greenhouse gases such as nitrous oxide (laughing gas) and chlorofluorocarbons (that have been used as refrigerants). These activities include burning of fossil fuels, clearing and not replenishing forests and other vegetation that absorb carbon dioxide, and producing more cattle and other livestock that emit methane. Fossil-fuel consumption contributes the largest share via electricity generation, transportation, and heating for residential and industrial purposes.

If excess greenhouse gases are not removed and their continued emissions are not curbed, the cumulative heating of the earth could lead to global harm. Our sister plant Venus demonstrates the hellish conditions that result from too much greenhouse effect. Venus's atmosphere has a carbon dioxide concentration of about 97 percent, resulting in an average surface temperature of 467° Celsius (872° Fahrenheit). Moreover, its atmosphere has a surface pressure approximately ninety times greater than the earth's. These conditions can melt lead and crush space probes that try to explore Venus.

What is the difference between global warming and climate change?

In 1896, Svante Arrhenius, a Swedish scientist, proposed that changes in carbon dioxide and water vapor could alter the earth's temperature. He predicted that as carbon dioxide and other greenhouse gas concentrations increase, the globe's average temperature would go up. His and other researchers' data have backed this theory. But the earth is a complex system. Feedback mechanisms both natural and human induced could cause changes in regional climates that are difficult, if not impossible, to predict. Like meteorologists forecasting probable changes in the weather, climate scientists can forecast probable changes in climate by developing more advanced techniques of computer modeling of the earth's complex systems, including the atmosphere, land, and oceans.

Because even scientists themselves are struggling with understanding these systems, it is no surprise that politicians and the general public tend to be confused about this subject. The terms used to describe what is going on further confound people. When the subject entered public discourse in the 1980s and 1990s, it was usually described as "global warming." This description had the benefit of being readily visualized or literally felt. During those decades, people could feel that hot months of the year were generally getting hotter. Many parts of the world were registering record-breaking heat for several of the years in that time period.

But the term "global warming" failed to embody the more profound consequences of the rising temperatures. Thus, climate scientists prefer the term "climate change." The earth, however, does not have one climate but, rather, regions of the earth can vary considerably in their climatic effects. And within regions, places can experience microclimates depending on the geographical conditions. This variance has led to many people wondering whether there will be "winners and losers" in terms of the changing climates. While some regions may experience benefits in terms of, for example, growing seasons for agriculture, people are interconnected through migration patterns and markets trading in food and other commodities. So, what happens in one region will likely have far-reaching international effects.

What are the observed and forecasted effects of climate change?

Global average surface temperatures increased about .6° Celsius (one degree Fahrenheit) over the twentieth century. While this increase may seem small, it has had demonstrable effects, and the average temperature continues to rise as concentrations of greenhouse gases increase. One noticeable effect is a rise in sea level. Heated water expands. In addition, the melting of land ice, such as in Greenland and Antarctica, has further raised sea levels. Another effect has been the thawing of permafrost in northern climes such as Alaska, Canada, and Siberia. One of the concerns of this effect is the unlocking of methane, a greenhouse gas, trapped in the permafrost. This would likely speed up the vicious cycle of more global warming.

The seas absorb some carbon dioxide, the primary greenhouse gas, from the atmosphere. Having more atmospheric carbon dioxide tends to increase the amount absorbed in the seas up to a saturation threshold at which the seas cannot absorb anymore carbon dioxide. This effect should concern us because absorbed carbon dioxide in the seas results in turning the water more acidic. This acid has bleached coral reefs and consequently has killed living coral. These major breeding grounds for a multitude of sea life are thus threatened by human-induced climate change. The world is already witnessing the shrinking of Arctic sea ice. This increasing loss has pushed polar bears, for example, to the brink of extinction. Within the next couple of decades, polar bears may exist only in zoos.

Forecasts indicate that the regional and global effects of climate change could become much worse. Over the course of this century, global average temperatures could rise between 1.4 to 5.8° Celsius (2.5–10.4° Fahrenheit), according to the consensus reached by the International Panel on Climate Change. By century's end, the average surface temperature could become 10 to 40 percent greater than today. One can tell by the relatively wide temperature range that this is a science of forecasting, not prediction. While the uncertainties are large, the trend lines point to even more drastic effects. For instance, many island-nations could become totally submerged. Rising sea levels would also likely flood coastal cities. Several of these cities are the world's megacities with more than 10 million residents each. Submergence and flooding would drive people to flee these areas, resulting in massive refugee crises. This migration would strain resources in other areas.

Another forecasted effect is that wet regions will become even wetter with increased rainfalls and snowfalls, and dry regions will become even drier with more droughts. In the former regions, more flooding is forecasted, and in the latter, more heat waves and wildfires would likely occur. One of the biggest worries is that the earth's climatic systems are approaching tipping points at which the effects become amplified and increase exponentially. Corrective natural and manmade mechanisms may eventually swing the systems back into a more desirable range. But the time required to make this correction depends on how long the greenhouse gases stay in the atmosphere. Unfortunately, the typical residence time of a carbon dioxide molecule in the atmosphere is more than a century. Consequently, even if people stopped burning fossil fuels today and did not remove much of the excess carbon dioxide, the amount of carbon dioxide already in the atmosphere will have an inertial effect, driving climate change.

What can people do to reverse excess global warming?

Essentially, people can take three actions: reduce the amount of greenhouse gases pumped into the atmosphere; increase the absorption of these gases—especially carbon dioxide; and decrease the amount of sunlight penetrating the atmosphere. For the first action, people can take a number of steps. One of the most important efforts is to decrease the amount of fossil fuels burned for transportation, electricity generation, and heating. The type of fossil fuel used also matters. In particular, the higher the carbon content of the fuel, the greater amount of carbon dioxide emitted when the fuel is burned. From highest to lowest carbon content, the main fossil fuels are coal, petroleum, and natural gas. So, substituting natural gas for coal could significantly reduce carbon dioxide emissions. A greater positive effect can occur by replacing fossil fuels with solar, wind, hydro, and nuclear power sources, which have very low greenhouse gas emissions associated with their lifecycles. (The role of nuclear power will be discussed more fully later.) Sequestering carbon dioxide emissions from fossil-fuel power plants may eventually become widely used. But carbon sequestration so far is employed in only a few places, such as in the Sleipner oil field operated by Statoil in the North Sea and in Beulah, North Dakota, at a coal-fueled synthetic natural gas plant.

Another step is to reduce the per capita emissions of fossil fuels. That is, each person would consume less of these fuels. But the pattern over the industrialized age has been that nations have consumed prodigious amounts of fossil fuels on the road to becoming more developed. The challenge is to figure out how to help the developing nations leapfrog the intensive fossil-fuel burning stage. A related challenge is to have the world's population reach a sustainable level. The global population is proceeding on a trajectory toward at least 9 billion by mid-century. The present population of around 6.8 billion is already on a nonsustainable path, consuming natural resources at a far greater rate than can be replenished. Imagine what business-as-usual would be like with 9 billion people.

Energy use has been historically correlated with population development. That is, greater use of energy has generally helped increase people's life spans and reduce infant mortality. This has gone hand-in-hand with a better educated populace, which has resulted in a leveling off, and in many developed countries a decline, of population as women tend to have fewer births. Another way to reduce greenhouse gas emissions is by people eating foods that are lower on the food chain. In particular, by eating less meat, people would lower the demand for livestock and thus decrease the amount of methane released by these animals and reduce the deforestation that often accompanies clearing land for cattle.

The second area of action involves taking carbon dioxide and other greenhouse gases out of the atmosphere. Reforestation, for example, would sequester carbon dioxide in trees. Stimulating the growth of other plants could also help. Commercializing fuel production from algae, switch grass, and other nonfood plants, and replacing fossil fuels with this biomass fuel, may eventually result in net carbon reduction from the atmosphere.

The final action area is known as geo-engineering. People could purposely pump reflective materials such as sulfates into the upper atmosphere. Other methods include brightening clouds by spraying tiny droplets of seawater into low-lying clouds to make them more reflective to block more sunlight from reaching the earth's surface and deploying sun shields made of trillions of tiny disks. While these geo-engineering techniques would most likely produce a global cooling effect, they do nothing to reduce the amounts of greenhouse gases. A concern is that they would encourage people to continue consuming more fossil fuels. And as soon as the reflective methods are not employed, the full effects of global warming would bear down on the earth. Moreover, these methods do nothing to alleviate the acidification of the seas. Nonetheless, if the world becomes more desperate for solutions, people may clamor for geo-engineering. Thus, serious research and development is needed to weigh the benefits and risks.

Why don't nuclear plants emit greenhouse gases?

Nuclear power plants do not combust fuels that release greenhouse gases as by-products. The by-products of the fission reaction are radioactive fission products that stay trapped inside the fuel assembly and, even if released as the result of an accident, would not contribute to the greenhouse effect. In contrast, power plants that burn carbon compounds such as coal, oil, or natural gas produce carbon dioxide during combustion with oxygen.

Why does the nuclear fuel cycle emit some greenhouse gases?

Although nuclear power plants do not emit greenhouse gases, other parts of the nuclear fuel cycle do. Starting with the mining of uranium, the mining equipment used to extract uranium from the ground and the trucks to transport the raw ore consume fossil fuels. After the milling process separates uranium ore concentrate from the other extracted mining material, trucks usually transport this concentrate to chemical-conversion plants. If the uranium requires enrichment, then there is further transit to an enrichment plant. The conversion and enrichment plants may consume a significant portion of the electricity they obtain from fossil-fuel sources, depending on the mix of electricity generation. This consumption of electricity can vary widely. For instance, the gaseous diffusion-enrichment plant still operating in Paducah, Kentucky, is very energy inefficient as compared to a modern gaseous centrifuge-enrichment plant and it uses a relatively large portion of the electricity it obtains from fossil-fuel sources. The United States Enrichment Corporation, the owner of the Paducah plant, is developing and trying to commercialize the American Centrifuge Plant, which, if successful, would be one of the most efficient enrichment plants in the world. Thus, deploying more efficient enrichment methods will reduce the already proportionally small amount of greenhouse gases emitted during the nuclear fuel cycle. Using equipment that is not powered by fossil fuels and using alternative-energy vehicles to mine uranium and transport nuclear fuel would further reduce these emissions.

How helpful have nuclear power plants been in preventing more greenhouse gases from being emitted from coal and natural gas plants?

At the end of 2009, the world's commercial nuclear reactors had a combined electrical power rating of about 370 GWe, or 370 billion watts of electrical power. In terms of the actual electrical energy generated, these reactors produced approximately 2,600 billion kilowatt-hours. This was about 15 percent of world electricity use. If these reactors were replaced with coal power plants, which generated about 41 percent of the world's electrical energy in 2009, about 4,000 million metric tons of carbon dioxide would be emitted in addition to the carbon dioxide that the coal plants were already producing. If natural gas plants replaced the world's nuclear plants, the additional increase in emissions of approximately 2,000 million metric tons of carbon dioxide would be about half of what it would be with coal plants.

How many additional nuclear plants would be needed to make a further major reduction in greenhouse gas emissions?

The "wedge" model can help us consider how many additional nuclear plants would be needed to make a further, significant reduction in greenhouse gas emissions. This model was published in *Science* in 2004, in an article by Stephen Pacala and Robert Socolow of Princeton University. In this groundbreaking study, the authors did the math on what contributions fifteen different energy technologies and practices can make toward reducing greenhouse gases. Rather than concerned with simply reducing emissions, they set the goal of flattening annual emissions to 2004 levels by mid-century. In 2004, global human activity released about 7 billion tons of carbon. Projecting future emissions based on business as usual practices, Pacala and Socolow predicted that annual emissions would at least double to 14 billion tons of carbon by 2054, fifty years from their starting point. No technology can singlehandedly be deployed in this time frame to displace all of these emissions. To make the problem more manageable, but still challenging, Pacala and Socolow employed the technique of divide-and-conquer. They split the fifty-year emissions into seven slices, or what they termed "wedges." Each wedge would represent an increase of 1 billion tons of carbon in annual emissions.

Fully deploying any wedge would require substantial investment. No wedge is easy to accomplish. For example, to fill a wedge with improvements in vehicle efficiency would require increasing the fuel economy for 2 billion cars from 30 to 60 miles per U.S. gallon (12.75–22.51 kilometers per liter). To accomplish this, from 2010 to 2054, every year about 44 million highly fuel-efficient cars would have to be built and displace less efficient cars. In comparison, the world made about 52 million cars in 2009. So, the challenge is clear and difficult; practically every manufactured car must reach very high fuel efficiency standards.

Another wedge would involve capturing and storing the carbon dioxide from 800 gigawatts (GW) of coal plants or 1,600 GW of natural gas plants. (A gigawatt equals 1 billion watts, or enough electricity to power about 1 million homes in the United States.) This wedge would equate to about 1,000 large coal plants or about 2,000 natural gas plants. But large-scale deployment of carbon sequestration from these plants is likely decades away. For biomass fuel to displace enough fossil fuel to fill a wedge, the world would have to add about 100 times the current Brazilian ethanol production, or approximately one-sixth of the world's cropland.

Concerning nuclear power's role in filling a wedge, Pacala and Scolow calculated that 700 GW of additional nuclear power would be required, or about double the current capacity. The 700 GW would translate into 700 large reactors of 1,000 megawatts each. Many of the newer generation reactors have higher power ratings of up to 1,600 megawatts; nonetheless, the 700 reactors provide an estimate. In addition to these reactors, almost all of the current fleet of about 370 reactors would have to be replaced by mid-century. Consequently, filling this wedge would require building about 1,000 reactors of 1,000 megawatts average size. This build rate from 2010 to 2054 would equate to about two new large reactors connected to the grid every month. Of course, fewer larger power rating reactors would have to be built. The largest power rating reactor is the 1,600-megawatt European Pressurized Reactor. If only these were built, the connection rate to the grid would have to be one of these about every three weeks to reduce greenhouse gas emissions by one-seventh of business-as-usual projections.

Have nuclear power plants ever been built as fast as would be needed to make another major reduction in greenhouse gas emissions?

According to the Keystone Center's 2007 report, the 1980s were the decade with the largest rate of nuclear growth. About 20 GWe were added on average every year. This is equivalent to connecting one 1,000-MWe reactor to the grid every two and half weeks—close to the build rate that would have to be done for nuclear power to make a significant further reduction in carbon dioxide emissions. But the build rate would have to be sustained over many decades. Notably, the construction rate fell off substantially in the 1990s with only the equivalent of forty-four large reactors being deployed.

Will global warming actually reduce the ability to use nuclear power plants to their full capacity?

Even if all the reactors to fill a wedge could be built, they may not be able to operate at full power because of global warming. Nuclear power reactors need large reservoirs of cooling water from seas, lakes, rivers, or man-made cooling ponds. The reactors dump waste heat into these reservoirs. Without enough relatively cool water to receive this heat, reactors could overheat. But these bodies of water would all experience increases in average temperatures because of global warming. Hot reservoirs can harm the animal life in them. Environmental regulations, consequently, are designed to protect this life. The hotter the reservoir, the less waste heat it can receive from a reactor. Thus, the reactor operator would have to throttle back on the power output, reducing the amount of electricity generated. To be fair to nuclear plants, large power plants powered by coal and other fossil fuels would also require reservoirs for their waste heat.

Should nuclear power be considered a "clean" energy source for climate-change agreements among nations?

During the past two decades, a contentious debate has been fought between those who believe that nuclear energy offers clean energy and those who believe that this energy source has too many risks. To review, energy is "clean" in the sense of not emitting greenhouse gases and other atmospheric pollutants such as sulfur dioxide, which contributes to acid rain, and nitrous oxides, which can harm respiratory systems. The risks inherent with nuclear energy include safety, proliferation, and waste disposal. The major forums for this debate have been the many international conferences dealing with climate change. In the late 1990s and early 2000s, the main arena was the Kyoto Protocol. This protocol to the United Nations Framework Convention on Climate Change called for reductions in four greenhouse gases: carbon dioxide, methane, nitrous oxide, and sulfur hexafluoride. The protocol required legally binding reductions only from the so-called Annex I countries—those that are considered industrialized and developed.

While the industrialized countries agreed to reduce their combined greenhouse gas emissions by 5.2 percent compared to 1990 levels, the amount of required reduction varies by country. The individual reduction assessments ranged from 8 percent for the European Union countries collectively, to 7 percent for the United States, 6 percent for Japan, and zero percent for Russia. Because Russia had only just emerged from the breakup of the Soviet Union by 1991, and had had a relatively weak economy in 1990, the basis year for reductions, it was not required to make further reductions below the 1990 level. Although the Clinton administration supported the protocol during the negotiations in the late 1990s, the subsequent Bush administration decided against asking the U.S. Senate for its advice and consent on ratification of the protocol. Bush administration officials objected to what they considered to be too deep cuts in U.S. emissions—cuts they thought would harm the American economy. They were also opposed to the protocol's exclusion from binding reductions such major gas-emitting countries as China and India in the developing world. Despite the United States' declining to join the protocol, 183 countries—a vast majority of the world— did ratify the protocol.

This protocol established the Clean Development Mechanism (CDM) as a way for developed countries to invest in no- or low-carbon emissions energy projects in the developing world. Lobbying for and against including nuclear power in the CDM has been intense. Consensus had not been reached on this issue in the lead-up to the negotiations on a post-Kyoto agreement. Thus, nuclear power was not deemed eligible for inclusion in the CDM. But at the round of negotiations in Copenhagen, Denmark, in December 2009, three options were considered: (1) continue the status quo exclusion, (2) prohibit Annex I countries from receiving carbon-reduction credits from nuclear power but open the door for possible credits in non-Annex I countries, and (3) establish a start date of January 1, 2008, so that nuclear plants deployed after this date may be eligible for carbon credits. At the December 2010 international round of climate change talks in Cancún, Mexico, no decision was reached to include nuclear power in the CDM.

What are the differing views among environmentalists on nuclear power?

Despite the longstanding skepticism, or even hostility, of many environmental watchdog groups toward nuclear energy, some prominent environmentalists have recently come out as strong supporters of expanded use of nuclear power plants. Of these supporters, Patrick Moore has captured significant attention because he was a founder of Greenpeace, an environmental organization that still maintains an anti-nuclear stance. Moore writes about how, in the 1970s, when he helped found Greenpeace, he equated nuclear energy with "a nuclear holocaust," but more than thirty years later his views have changed. In the *Washington Post* on April 16, 2006, Moore argued that "nuclear energy may just be the energy source that can save our planet from another possible disaster: catastrophic climate change." Far from joining him in his change of heart, many Greenpeace employees have considered Moore a turncoat. They point to his work as a corporate consultant since 1991, and they are especially alarmed at his work for the Clean and Safe Energy Coalition, which is affiliated with the Nuclear Energy Institute (NEI), the nuclear industry's lobbying firm. The coalition has received funding from the public relations firm Hill and Knowlton, which had an $8 million account with NEI. In October 2008, Greenpeace stated that Moore "exploits long gone ties with Greenpeace to sell himself as a speaker and pro-corporate spokesperson, usually taking positions that Greenpeace opposes." As part of this campaign, Moore has teamed up with former George W. Bush administration Environmental Protection Agency head Christine Todd Whitman. While Moore's promotion of nuclear power would lead one to believe that he is convinced that climate change is due to human activities, remarkably he doubts that "global warming is caused by humans, but it is likely enough that the world should turn to nuclear power."

Other prominent pro-nuclear environmentalists are far more concerned about human-induced climate change. British scientist James Lovelock—called by *Rolling Stone* "the Prophet of Climate Change"—fears that climate change is irreversible and by the end of the century it could claim upwards of 6 billion lives. He is most famous for devising the Gaia Hypothesis, a view that earth is a self-regulating system. Warning, in a May 24, 2004, op-ed in the *Independent*, that there is "no time to experiment with visionary energy sources," Lovelock believes that countries have "to use nuclear—the one safe, available, energy source—now or suffer the pain soon to be inflicted by our outraged planet."

While Moore and Lovelock have grabbed news-media attention, perhaps the most effective pro-nuclear environmentalists are those who know how to make the case for nuclear power in the halls of political power. Politically plugged-in Jonathan Lash, the president of the World Resources Institute, sees nuclear power as "a necessary evil" and as part of a multipronged strategy to combat climate change. Understanding that environmentalists alone cannot turn the tide on climate change, he has bridged the divide between many environmental groups and big industries by helping found the United States Climate Action Partnership (USCAP). This organization has married the idealism of environmentalists with the profit-seeking motive of business. More than a dozen major companies, such as General Electric, Alcoa, Duke Energy, BP America, DuPont, Exelon, and NRG Energy, have joined USCAP. In addition to the World Resources Institute, USCAP's environmental groups include the Natural Resources Defense Council, the Nature Conservancy, the National Wildlife Federation, and the Pew Center on Global Climate Change. The organization's monetary and market power contains revenues of more than $1.7 trillion and a workforce of some 2 million.

4
PROLIFERATION

What is nuclear proliferation?

The term "nuclear proliferation" refers to two types of activities: the acquisition of nuclear weapons by countries that do not have them; and the increase of nuclear arsenals in countries that already have nuclear weapons. The first activity is known as "horizontal proliferation" because of its spreading to states without nuclear arms; and the second activity is known as "vertical proliferation" because of its building up of warheads in the manner of bricks being added to a tower. An example of horizontal proliferation is North Korea's development of a small nuclear arsenal. This action may spark other Asian states such as Japan and South Korea to contemplate acquiring and eventually building nuclear weapons. The exemplar of vertical proliferation was the massive buildup of American and Soviet arsenals during the Cold War. While the United States and Russia are thankfully decreasing their arsenals, India and Pakistan are building up their nuclear arsenals. The South Asian arms race is worrying because it is in an unstable political region, with political coups in Pakistan, frequent terrorism, and the presence of some terrorist groups that would want to acquire nuclear weapons.

Which countries have developed nuclear weapons, and how did they do it?

The spread and growth of nuclear arsenals are linked in a political chain reaction. That is, proliferation has spurred further proliferation.

The first nuclear-weapons program ignited in the cauldron of the Second World War. Facing the threat of a potentially nuclear-armed Nazi Germany, the United States rushed to build nuclear weapons in the secret Manhattan Project. This project produced two types of weapons: the gun-type bomb that was detonated above Hiroshima on August 6, 1945; and the implosion-type bomb that was detonated above Nagasaki on August 9, 1945. Both bombs destroyed the cores of these cities and in total killed a couple of hundred thousand people. These bombings marked the end of the Second World War. The Cold War between the Soviet Union and the United States immediately followed.

Seeking to counter the U.S. monopoly on possession of nuclear weapons, Soviet leader Joseph Stalin ordered a crash program to acquire these weapons. The Soviet Union's nuclear bomb program benefited tremendously from spies such as Klaus Fuchs and Ted Hall, who had worked in the Manhattan Project. Spying had given the Soviets the details of the Nagasaki bomb so that Soviet weapon scientists could choose to make that same type of bomb and guarantee a successful first nuclear test. Failure could have meant exile to the Gulag prison system or even worse, a bullet to the head. KGB Chief Lavrenti Beria ruthlessly oversaw the Soviet Union's nuclear weapons program. Beria and Stalin's iron discipline paid off with a successful test in August 1949. The quickness of the Soviet's bomb program shocked U.S. President Harry

Truman. This test helped convince him to support the buildup of the U.S. arsenal. Fears that the Soviets would develop even more powerful nuclear bombs led the United States to accelerate research and development of so-called hydrogen bombs, or thermonuclear weapons.

While the United States and the Soviet Union became locked in a potentially deadly nuclear arms race, Great Britain and France sought their own nuclear arsenals. Great Britain became the third nuclear-weapon state with its test in October 1953, and France struggled to follow suit in February 1960. As major victors of the Second World War, both of these countries felt compelled to acquire the ultimate weapon as a means of helping to ensure great-power status. While Britain and the United States had and still do have a close defense relationship that has included sharing of some nuclear weapon systems such as the Trident missile for submarines, France was further motivated to build its own nuclear weapons because of its perceived need to exert its independence from the United States.

The birth of Communist China in 1949 brought Chairman Mao Zedong to power, and as a result the world's most populous nation became an ideological enemy of the United States. The United States had backed the nationalist Chinese, who fled to Taiwan. While Mao publicly derided U.S. nuclear weapons as "paper tigers," he felt threatened by these weapons and believed that China must acquire them to be immune from nuclear blackmail. In the 1950s, his weapon scientists received assistance from their Soviet counterparts. This proliferation aid continued until the Sino-Soviet split in 1962. Chinese weapon scientists were able to continue the program successfully. In October 1964, China became the fifth nuclear-weapon state with a powerful test explosion.

The Chinese test further convinced India that it needed nuclear weapons. An earlier stimulus was the 1962 Sino-Indo border war in which India lost some territory. India had been exploring a nuclear-weapons program since the 1950s, but the Chinese acquisition was a tipping point. Indian leaders sought to check the growing influence of China's power. The 1974 Indian nuclear test, which was dubbed "a peaceful nuclear explosive," resulted in a less than peaceful reaction: Pakistani leaders became resolved to acquire their own nuclear bombs. (Another ironic twist was the code name of the Indian test, "the Buddha smiles.") By the mid-1980s, Pakistan had nuclear-weapons capabilities, which were proven in a series of tests in May 1998, in response to a series of Indian tests earlier that month.

Sometimes direct nuclear threats are not necessary to stimulate a nuclear-weapons program. Perceived threats to a country's existence can stimulate such a program. Surrounded by hostile Arab states, Israel feared for its existence. In the 1960s, it began producing fissile material for nuclear weapons. Israeli leaders have adopted a policy that Israel would not be the first state to introduce nuclear weapons to the Middle East. This policy has been interpreted to mean that Israel will not openly deploy nuclear weapons, or even acknowledge its possession of such weapons, unless another state in the region does so. No other country in this region has developed nuclear weapons. But Iran has acquired a latent capability to do so.

North Korean leaders also fear existential threats to their regime. The Korean War from the early 1950s is not officially over; only an armistice has been signed. North Korea, often called the "hermit kingdom" because of its pariah status, is run by the Kim family dynasty—a cult of personality founded by Kim Il Sung, "the Great Leader," and taken over after his death in 1994 by his son Kim Jong Il, "the Dear Leader." The elder Kim started North Korea's nuclear program with assistance from the Soviet Union during the Cold War. North Korea has used a medium-size research reactor to produce weapons-grade plutonium. This reactor could generate about one bomb's worth of plutonium annually. North Korea detonated its first nuclear explosion in October 2006. While this explosion was widely considered less than optimal, if not almost a dud, by technical experts outside of North Korea, the hermit kingdom demonstrated a more potent nuclear device in May 2009. Since 2003, the United States has been working with China, Japan, Russia, and South Korea in the six-party talks to convince North Korea to give up its nuclear-weapons programs in exchange for security guarantees, U.S. diplomatic recognition, and economic assistance. But the six-party talks have suffered setbacks as North Korea has been undergoing major political changes. Kim Jong Il, who has reportedly experienced significant health problems in recent years, appears to value continued possession of nuclear weapons to preserve his family's leadership. North Korea has been in a leadership transition in which Kim Jong Il has been grooming his third son, Kim Jong Eun, for likely succession.

In sum, the eight known possessors of nuclear weapons are China, France, India, North Korea, Pakistan, Russia, the United Kingdom, and the United States; the ninth undeclared nuclear-arms possessor is Israel. Dozens of other states have the potential within several years to acquire nuclear weapons if they decide to break free of their international commitments and harness their civilian nuclear infrastructure.

How many nuclear weapons do the nuclear-armed countries have?

Nuclear-armed states generally do not publish the numbers of nuclear weapons they have. The one big exception is the requirement under the Strategic Arms Reduction Treaty (START) between Russia and the United States for both countries to declare the numbers of their strategic nuclear-weapon systems. Under START, however, the weapons-delivery systems are the focus of the treaty because these items are relatively easy to count and thus verify. The three types of strategic delivery systems are intercontinental ballistic missiles (ICBMs) based on land either in stationary silos or mobile launchers, submarine launched ballistic missiles (SLBMs) based on submarines, and long-range bombers that can carry cruise missiles or bombs. Individual warheads, on the other hand, are easier to hide. No treaty has yet to require strict verification of nuclear warheads.

START and other Russia–U.S. arms control treaties have defined strategic and nonstrategic weapon systems. Because of the world-spanning distance between these two nuclear-armed rivals, strategic systems were defined as long-range weapons such as those described above. Nonstrategic systems in the American–Russian context are shorter range weapons. Russia has altered its nuclear doctrine since the end of the Cold War to allow use of nonstrategic nuclear weapons to respond to conventional attacks. The Russian conventionally armed military is much weaker than NATO's conventional forces. Thus, nonstrategic nuclear weapons are often termed tactical weapons because of these potential battlefield deployments. Many analysts argue that any nuclear weapon is a strategic system in the sense that it affects a country's political decision making and thus its overarching national

TABLE 4.1: *Nuclear-armed countries.*

Country	Estimated number of weapons (end of 2009)
China	100–200
France	350
India	Up to 100
Israel	75–100
North Korea	Up to 10
Pakistan	70–90
Russia	4,600
United Kingdom	225 (actual)
United States	5,113 (actual)

strategy. Certainly, within the South Asian context in which India and Pakistan border each other, any nuclear weapon, no matter what range it has, is a strategic weapon.

Because it is important for the public and policymakers to know the status of nuclear arsenals, several independent, nongovernmental organizations have made estimates. These organizations include the Arms Control Association, the Center for Defense Information, the Federation of American Scientists, the International Institute for Strategic Studies, the Natural Resources Defense Council, and the Stockholm Peace Research Institute. In early May 2010, U.S. Secretary of State Hillary Clinton announced that the United States on September 30, 2009, possessed 5,113 warheads in the deployed and active reserve stockpiles and that several thousand more warheads await dismantlement. In late May, the British government announced that it has up to 225 warheads with about 160 maximum in a deployed status. The United States and Russia possess more than 90 percent of the world's nuclear weapons, as shown in table 4.1, which lists the best estimates and the two official announcements of the number of these weapons. This table does not include several thousand warheads in Russia and the United States that have been retired or waiting to be dismantled. According to Hans Kristensen and Robert S. Norris, who wrote the Nuclear Notebook for the *Bulletin of the Atomic Scientists*, the total global inventory is about 22,400 warheads.

How much weapons-usable fissile material is available worldwide, and where is it located?

While thousands of nuclear weapons are worrisome, the growing stockpiles of weapons-usable fissile material pose increasing threats of proliferation and nuclear terrorism. Two types of fissile material present the greatest concern: highly enriched uranium (HEU) and plutonium. According to the International Panel on Fissile Materials, as of the end of 2009, worldwide there are more than 1,500 metric tons of highly enriched uranium in the world, and about 500 metric tons of separated plutonium (outside of the protection barrier of highly radioactive fission products). This amount of fissile material could fuel tens of thousands of nuclear bombs. Security experts have pointed out that the uncertainties in the data are large, especially for the material in Russia.

Both highly enriched uranium and plutonium are used in military and civilian applications. In military applications other than weapons purposes, weapons-grade uranium and plutonium power nuclear explosives and fuel many warships, including submarines and aircraft carriers. In the civilian applications, some commercial reactors are fueled with recycled reactor-grade plutonium, which is weapons usable. (All commercial power reactors derive part of their power from the fission of plutonium that was generated inside the reactor.) Also, dozens of research and test reactors still use highly enriched uranium. Certain kinds of these reactors employ HEU as target material to make medical isotopes. In addition, Russian icebreakers have used weapons-usable uranium in reactors to generate electrical power and propulsion. Security programs are working to phase out the use of highly enriched uranium in civilian applications. While one of the main impediments had been finding non-weapons-usable substitute fuels and target materials, scientific and engineering breakthroughs have been eroding this technical hurdle. The remaining roadblocks are a lack of political will to commit to eliminating civilian HEU and the high financial cost of finding alternatives to the existing technologies. While the eventual elimination of civilian HEU shows much promise, the phase-out of recycled plutonium fuels is far less likely because a few major nuclear power countries such as France, India, Japan, and Russia remain committed to this practice.

Russia and the United States possess most of the world's fissile material. While these two countries have declared several hundred tons of HEU as excess to defense needs, and have been converting this excess to non-weapons-usable forms, they still have reserved several hundred tons for weapons purposes and naval reactor fuel. The declared excess material has been fueling a significant portion of commercial reactors. For example, the United States has been purchasing the converted Russian weapons-usable uranium to fuel about half of the 104 U.S. reactors or about 10 percent of U.S. electricity. Thus, one in every ten light bulbs in the United States is lit by uranium from dismantled Russian warheads. This program is scheduled to continue until the end of 2013. The United States and Russia have had some discussions but no agreement yet on extending this program to convert additional weapons-usable uranium into reactor fuel.

Most of the nuclear-armed states have ceased production of military plutonium. Russia, the United States, and the United Kingdom have declared more than ninety tons as excess to defense needs. Of the original five nuclear-weapon states, only China has not made an official declaration of stopping production, but it is generally believed to have stopped. However, India and Pakistan have continued to produce military plutonium because they are competing in a nuclear-arms race. And North Korea has taken steps to restart its plutonium production capability and may complete additional plutonium-production reactors.

While the growth of Indian, North Korean, and Pakistani military plutonium stockpiles poses a proliferation threat, the increase of civilian plutonium presents a latent security concern. Nuclear weapon states such as France, Russia, and the United Kingdom that separate plutonium from spent civilian fuel obviously do not pose a horizontal proliferation threat because they already have nuclear weapons. Japan is the only non-nuclear-weapon state with reprocessing facilities to separate plutonium from spent fuel. Japan's growing plutonium stockpile and its future production capacity have worried China. As of 2010, according to the International Panel on Fissile Materials, the total stockpile of separated plutonium for civilian use is roughly equivalent to the military stockpile at about 250 metric tons each, but the civilian stockpile is increasing faster than the military stockpile. This material could power thousands of nuclear warheads.

Has a country ever completely dismantled or given up its nuclear arsenal?

South Africa stands out as the only country to dismantle completely a nuclear arsenal. During the 1970s, South African leaders felt under threat because of its increasingly isolated status owing to its apartheid regime. They believed that nuclear weapons may enhance their security but they may not have ever seriously considered using nuclear weapons. As Mitchell Reiss has observed in *Bridled Ambition*, South African leaders devised the option of detonating a nuclear weapon as a political signaling device. In this scenario, the nuclear explosion would likely have occurred away from population centers.

South Africa acquired assistance from West Germany in building a uranium enrichment plant. This plant provided the highly enriched uranium for six gun-type nuclear bombs. South Africa was in the process of building a seventh nuclear bomb before President F. W. de Klerk decided in the late 1980s to eliminate the weapons program. De Klerk knew that the apartheid regime was destined for the trashcan of history. That is, the apartheid system of government had forced the majority black African population into a subservient role. During the final decade of this regime in the 1980s, international sanctions were becoming more effective in helping to convince South Africa's leaders to end their discriminatory rule. After South Africa dismantled its nuclear weapons, it invited inspectors from the International Atomic Energy Agency to confirm the dismantlement. But South Africa continues to possess a stockpile of weapons-usable uranium, which is devoted to civilian use.

Inherited nuclear weapons in new states can pose a challenge. The breakup of the Soviet Union in 1991 created fifteen independent states from the behemoth Soviet state. While Russia was the designated successor state for possession of the nuclear weapons, Belarus, Kazakhstan, and Ukraine still had Soviet nuclear weapons on their soil. The Russian government wanted those weapons to be given to Russia. But having these weapons provided considerable leverage for these three new states. Creative diplomacy and various financial incentives helped convince the governments of these states to relinquish these nuclear arms and join the international community as non-nuclear-weapon states.

What is the nonproliferation regime?

Designed to stop the spread of nuclear weapons to additional countries, the nonproliferation regime consists of the Non-Proliferation Treaty, international institutions such as the International Atomic Energy Agency (IAEA) and the Nuclear Suppliers Group, and bilateral nuclear cooperation agreements in which client-states agree to accept safeguards and monitoring of their peaceful nuclear programs in order to receive technologies and assistance for these programs. The United Nations Security Council is responsible for enforcing this system. In past cases of proliferation concern, the Security Council has passed resolutions that call on violators to come back into compliance with the rules and that at times impose economic, military, or political sanctions on violators.

What is the nuclear Non-Proliferation Treaty?

In the early 1960s, U.S. President John F. Kennedy warned that the trend toward further proliferation could result in fifteen or more new nuclear-armed states by the 1970s. When he spoke, at least a couple of dozen states were exploring nuclear-weapons programs. Even before his speech, diplomatic steps were being taken to halt the further spread of these programs. In the late 1950s, Ireland initiated the process by proposing to the United Nations a resolution on the "nondissemination of nuclear weapons." Irish efforts through 1961 helped pave the way for an eventual treaty on nonproliferation. But the United States was the key state. Without its support, there was no hope for such a treaty.

In 1964, President Lyndon Baines Johnson approved the formation of the Gilpatric Committee to assess what the United States could do to stop nuclear proliferation. The immediate stimulus for forming the committee was China's October 1964 nuclear test. Chaired by Roswell Gilpatric, former Deputy Secretary for Secretary of Defense Robert McNamara, this committee examined the options of accepting proliferation as unstoppable, acknowledging proliferation to be a problem worth trying to stop but not at the expense of more pressing concerns, making a concerted effort to stop proliferation as a top priority, and taking aggressive action, even preemptive military attacks, to prevent proliferation. In a May 14, 2003, seminar at the Massachusetts Institute of Technology, Francis Gavin of the University of Texas uncovered three paradoxes in the committee's report: (1) "the more effort the United States put into counterproliferation, the more valuable nuclear weapons appeared to smaller powers for use as political bargaining chips"; (2) "effective counterproliferation could only be achieved with the cooperation of the Soviet nemesis"; and (3) "the goals of counterproliferation were often incompatible with the American nuclear posture"—that is, American pledges to provide nuclear deterrence to allies could undermine U.S. efforts to reduce the perceived value of these weapons. The committee recommended that the United States pursue close cooperation with the Soviet Union to stem the further spread of nuclear weapons. The report's findings remained classified for three decades; nonetheless, its recommendations gave support to Johnson to pursue the Non-Proliferation Treaty.

Having entered into force in 1970, the Non-Proliferation Treaty (NPT) embodies three principles: preventing the spread of nuclear weapons to additional states, providing access to peaceful nuclear energy to states that abide by the rules of the treaty, and pursuing nuclear disarmament as well as general and complete disarmament. Because five states had openly acquired nuclear explosives before the treaty was opened for signatures, they received special status as nuclear-weapon states. The treaty defines a "nuclear-weapon state" as a state that exploded a nuclear device before January 1, 1967. All other states—even if they have not joined the treaty— are defined as nonnuclear-weapon states. So, India, Israel, and Pakistan, which have never signed the NPT, are still considered nonnuclear-weapon states. North Korea joined the treaty in 1985 but left in January 2003, citing concerns about its security. Aside from the UN Charter, the NPT is the most universally adhered-to treaty, with 188 members. All but five of those members are nonnuclear-weapon states.

Often, the NPT is described as a grand bargain in which the nonnuclear-weapon states have pledged to not acquire nuclear explosives and to maintain adequate safeguards on their peaceful nuclear programs in exchange for the supplier states' providing access to peaceful nuclear technologies. Moreover, the nuclear-weapon states have pledged to pursue nuclear disarmament and a treaty on general and complete disarmament. Embedded within this bargain are some controversies that have vexed the full implementation of the NPT. First, although article IV of the treaty does point to an "inalienable right" to peaceful nuclear technologies contingent with the recipient's maintaining adequate safeguards, it does not explicitly mention access to the dual-use enrichment and reprocessing technologies. Some nonnuclear-weapon states, such as Brazil, Iran, and Japan, have interpreted this article to favor their acquisition of enrichment or reprocessing facilities. But these states differ in their adherence to safeguards. Brazil has opened its enrichment facility to IAEA inspection but resists implementing a more rigorous safeguards system; Japan, in comparison, has implemented the more rigorous set of safeguards; and Iran, by contrast, has been found to be in violation of its safeguards commitment. Some nonprolifera-tion experts and politicians have called for preventing addi-tional nonnuclear-weapon states from acquiring these facilities while others have put forward criteria that would determine which states could qualify for such technologies.

Another related controversy is how to put up barriers to withdrawal from the treaty and what to do if a state with-draws. A member state has the option under article X of the NPT to invoke its supreme national interests and leave the treaty after ninety days have elapsed. Although only North Korea has so far withdrawn, Iran may be next. If Iran with-draws, the worry is that many other states may follow, result-ing in the demise of the nonproliferation regime. Many experts have proposed that withdrawing states should be required, especially if they are in violation of their safeguards agreements, to open themselves to a special international inspection that would determine if any peaceful nuclear materials or technologies have been diverted to weapons purposes. An additional proposal is that withdrawing states that have violated the terms of the NPT should be required to return materials and technologies to the suppliers. Enforcing these proposals would be extremely challenging.

The other outstanding controversy is that there is no time-bound commitment for when nuclear disarmament should occur. In fact, the treaty refers only to "pursuit of nuclear dis-armament" and also in the context of "general and complete disarmament." Many believe that complete disarmament is an unrealistic goal. Nonetheless, even the serious pursuit could lead to greater security as long as the utmost attention is paid to ensuring that states feel secure during this process. U.S. President Barack Obama has recommitted the United States to pursuing nuclear disarmament, but he has cautioned that the United States will maintain a safe, secure, and reliable nuclear deterrent as long as other states have nuclear weap-ons. While this statement may seem to imply that nuclear disarmament can never be achieved, another way to look at it is that if all nuclear-armed states could agree to give up their nuclear arms, then the pathway to disarmament would look much more promising.

What is the International Atomic Energy Agency, and what role does it play in preventing proliferation?

The International Atomic Energy Agency is an international organization that is in the United Nations' family of agencies, but has a unique independent charter within that family. Born in 1957, the IAEA was a brainchild of U.S. President Dwight Eisenhower's famous "Atoms for Peace" speech on Decem-ber 8, 1953. He envisioned an atomic energy agency that would help ensure access to peaceful nuclear technologies. This promotional function of the agency has at times been a source of tension with the mandate to monitor peaceful nuclear programs and sound an alarm if states misuse these programs to build nuclear weapons. For example, Iran, the state that poses the greatest proliferation concern, has been the largest recipient of technical cooperation from the IAEA. In general, when the major developed states such as the United States and the United Kingdom have called for greater action to prevent proliferation, leading developing states such as Brazil and Egypt have indicated that the price to be paid is greater peaceful nuclear assistance. When the IAEA Board of Governors has determined that a nonnuclear-weapon state has violated its safeguards agreement, it is required to report that violation to the UN Security Council.

In addition to promoting peaceful nuclear applications and preventing proliferation, the IAEA is the leading international agency in developing safety standards and guidance documents for nuclear power. In more recent years, especially after the terrorist attacks on September 11, 2001, the IAEA has sought to build up its capabilities to offer member states assistance in securing nuclear materials and facilities that may be vulnerable to malicious people. But the IAEA makes it clear that safety and security are the states' responsibilities.

What are nuclear safeguards, and how have they evolved?

According to the IAEA, "safeguards are measures through which the IAEA seeks to verify that nuclear material is not diverted from peaceful uses." When effective, safeguards deter a state from diverting this material by increasing the probably of getting caught. So, it is not accurate to say that safeguards prevent proliferation. A state's decision to proliferate involves a complex set of political decisions weighing

domestic and international influences and technical requirements, and assessing whether it has or can acquire the human
and technological resources to make nuclear weapons. Safeguards, in effect, raise the cost of carrying out an act of proliferation. The safeguards system is also limited in its ability to
force a state to comply. States will benefit the most from this
system when they are assured that their neighbors are par-
ticipating fully and when they can make such assurances to
their neighbors.

States want to preserve their sovereignty and have a natural tendency to minimize international intrusion on their
activities. But because of the dual-use nature of peaceful
nuclear programs, states have a mutual interest to transcend
sovereignty and open these programs to inspection in order
to convince their neighbors, and especially rivals or enemies,
that they are not acquiring nuclear weapons. As long as every
state provides complete access, the safeguards system should
work effectively. However, many states continue to resist providing adequate access. States have also resisted providing
adequate financial support for the IAEA in carrying out its
safeguards mission. Consequently, the IAEA is constantly
cajoling states into being more cooperative and generous
with support.

Safeguards have evolved in reaction to crises. Although the
IAEA was formed in 1957, a couple of years elapsed before it
began to safeguard peaceful nuclear programs. In 1959, the
IAEA's Board of Governors approved the application of safeguards to a small research reactor that Japan was acquiring
from the United States. This act helped set a precedent for
reactor technology to be safeguarded. The next major step in
the 1960s was to apply safeguards to specific facilities, especially dual-use enrichment and reprocessing facilities. In 1970,

the entry into force of the NPT further strengthened the safeguards system. Specifically, article III in the NPT gives the IAEA authority to reach "comprehensive safeguards agreements" with all nonnuclear-weapon states party to the treaty. Most states presently belong to the comprehensive-safeguards agreement system, but there are still some laggards. Comprehensive safeguards, despite the word *comprehensive*, are not enough. A major loophole is that these safeguards apply only to a state's declared nuclear facilities and that the IAEA is not required to make a formal assessment of whether a state has any undeclared facilities. From the 1980s to 1991, Iraq exploited this loophole by placing undeclared facilities next to declared ones. IAEA inspectors had access to the latter but were denied access to the former. Nonetheless, the IAEA has the authority in its statute to demand access under the special inspection provision. But the IAEA Board of Governors has been reluctant to exercise this authority.

The revelation in 1991, after the First Gulf War, that Iraq was nearing acquisition of enough fissile material to make a nuclear bomb shocked the safeguards system. As a result, the IAEA Board authorized development of a more rigorous system. This Model Additional Protocol to Comprehensive Safeguards, or Additional Protocol, for short, seeks to change the mindset of IAEA inspections so that inspectors transform themselves from mere accountants verifying a state's declared materials and facilities to detectives investigating whether a state has any undeclared materials and facilities. The Additional Protocol also provides inspectors with greater access to the entire fuel cycle, including mining and milling activities. Moreover, it allows inspectors to access a suspected facility much faster than the previous agreement. In particular, if an inspector is at a site and has reason to suspect that something suspicious is happening at a facility on that site, he or she may request access within two hours after the request. If the inspector is off site, access should be provided within twenty-four hours. While about half the countries have signed the Additional Protocol, fewer have actually fully implemented it. More disconcerting is that nuclear suppliers have yet to agree to make the Additional Protocol a prerequisite for receiving nuclear materials and technologies. Egypt, for instance, has resisted adhering to the Additional Protocol because it feels it is an additional burden and it wants to maintain leverage on drawing attention to Israel's status as a non-NPT state.

Non-NPT states are not required to agree to comprehensive safeguards, but they have applied facility-specific safeguards to some peaceful nuclear facilities. Major nuclear suppliers have generally agreed to not provide nuclear materials and technologies to these states. However, India in 2008 was able to carve out a major exemption to this practice. The second Bush administration convinced the Nuclear Suppliers Group (which is described in more detail later) to relax its guidelines to allow India to access the international nuclear market in exchange for India's placing more, but not all, of its peaceful nuclear facilities under safeguards. This action has formed an additional double standard in that nonnuclear-weapon states are required to apply comprehensive safeguards while a non-NPT nuclear-armed state can access the market and expand its peaceful nuclear industry without relinquishing its nuclear weapons.

The original double standard is that between the official five NPT nuclear weapon states and the NPT non-nuclear weapon states. The former are not required to apply safeguards on all of their peaceful nuclear programs but may accept voluntary safeguards agreements on selected facilities and materials. Although the United States, for instance, has opened its peaceful facilities to the IAEA for inspection, the IAEA in practice has rarely done these inspections in order to conserve its limited resources and the fact that the United States as a nuclear weapon state has no incentive to divert fissile material from peaceful facilities. In sum, the present safeguards system is a patchwork: the majority of states have applied comprehensive safeguards; an increasing number of them are implementing the Additional Protocol; a few states are outside the NPT safeguards system and have applied some facility specific safeguards; and the five nuclear weapon states may accept voluntary arrangements.

Are nuclear safeguards effective?

Given the patchwork nature of the safeguards system, doubts have certainly arisen about its effectiveness. Even if there were no concerns about double standards, the comprehensive safeguards system and the companion Additional Protocol still have shortcomings. If one were to devise an ideal system, it would likely have continuous near-real-time monitoring of all materials and technologies in peaceful nuclear programs and wide-area environmental sampling to try to detect undeclared activities such as clandestine enrichment and reprocessing facilities. Of course, such a system would require substantially more resources than the IAEA currently has available.

Resource constraints have created a dysfunctional system, according to some prominent nonproliferation analysts such as Thomas Cochran and Henry Sokolski. When Mohamed ElBaradei was director-general of the IAEA, he warned that, without adequate resources, the agency risks becoming "hollowed out" and less than world class. In 2008, the Eminent Persons Panel, chaired by former Mexican President Ernesto Zedillo, called for a major expansion of the IAEA's budget to address the growing mismatch between the amounts of nuclear material requiring safeguards and monitoring and the amount of funding available to the agency to carry out this mission. The panel's report helped spur a small increase in the budget, but not nearly enough to close the gap.

The effectiveness of safeguards hinges on a credible system of detection and deterrence. In order to have confidence that the IAEA can deter diversion of nuclear material into weapons programs, states must believe that it is able to detect such diversion in a timely manner. The timeliness depends on how long a state would need to convert the material into weapons. This conversion time depends on the chemical form of the material. The ideal weapons-usable materials are pure weapons-grade uranium and plutonium metal. According to the IAEA, the time it takes to convert these materials into weapons is seven to ten days. The problem is that the IAEA has not been inspecting frequently enough to detect this potential diversion and conversion. In particular, the IAEA's timeliness detection goal for this material is one month. But the agency has usually not inspected nuclear facilities even that frequently. For other materials, the conversion times are longer. Highly enriched uranium and plutonium in oxide form require one to three weeks; HEU and plutonium embedded in irradiated or spent fuel require one to three months; and low-enriched uranium requires three months to one year. The IAEA's timeliness detection goals and periods between inspections are typically longer than these conversion times.

The type of nuclear facility also significantly affects the effectiveness of safeguards. For example, safeguarding declared reactors is relatively easy to do because, when visiting these reactors, it is possible to count discrete items such as spent fuel assemblies and the large and thus highly visible character of these items. In comparison, safeguarding fissile-material-handling facilities such as enrichment and reprocessing plants is much harder to do because these facilities are moving around large amounts of bulk materials over considerable lengths of piping and other pieces of equipment. So, such facilities tend to have materials unaccounted for, and thus raise the possibility of diversion of one or more bombs' worth of fissile material.

What is the Nuclear Suppliers Group?

The Nuclear Suppliers Group, or NSG, includes the major suppliers of nuclear technologies, but it does not include all the major suppliers of natural uranium. The forty-six countries in the NSG have banded together to develop guidelines to control the sale of sensitive technologies that could contribute to weapons programs. It is important to stress that these guidelines are not ironclad rules. But the NSG members have by and large adhered to these guidelines, which have evolved since the NSG was founded in the mid-1970s.

Because the 1974 Indian nuclear explosive test used plutonium produced from American and Canadian technology, the United States, Canada, and several allied countries sought to prohibit further misuse of related technologies. In particular, India had acquired the Cirus reactor from Canada. This reactor uses a special type of water called heavy water to control the nuclear reaction and cool the reactor's core. The United

States had provided heavy water for the reactor. Heavy-water reactors tend to be well suited to produce weapons-grade plutonium.

Canada and the United States leveraged the existing Zangger Committee to form the NSG. The Zangger Committee formed in 1971 among a group of major supplier states to determine how to effectively implement the Article III safeguards requirements in the Non-Proliferation Treaty, which had just entered into force the previous year. In 1974, the Zangger Committee published a trigger list of equipment that, if sold to a nonnuclear-weapon state, would trigger the requirement for implementing safeguards. The committee specified three conditions for supplying nonnuclear countries with equipment on the trigger list: (1) nuclear trade would not enable a state to develop nuclear explosives; (2) a recipient state would have to enact a safeguards agreement with the IAEA; and (3) any retransfer of equipment would require that the new recipient implement a comparable level of safeguards.

In 1978, the NSG published its first guidelines, following the path laid out by the Zangger Committee. In addition, the NSG's guidelines called for physical protection measures on nuclear materials and facilities and further strengthened retransfer provisions. Moreover, the NSG asked for formal government assurances that purchased equipment and materials would be subject to safeguards. But the initial set of guidelines did not explicitly call for safeguards on all nuclear activities in a country. For instance, a country could have a number of indigenously developed nuclear facilities and materials. While nonnuclear-weapon states that had signed the NPT would be required to place all declared facilities and materials under IAEA safeguards, those few states outside of the NPT, such as India, Israel, and Pakistan, were not required to do so. The question before the NSG was whether to allow these states to partake in nuclear commerce. A related issue was whether to allow commerce limited to activities specified in agreements that were made before the guidelines prohibited these activities. To address these issues, in 1992, the NSG further tightened its guidelines by specifying that nuclear commerce could take place only if a state had "full scope safeguards" on its entire nuclear program.

Still, the guidelines had safety and grandfather clauses that could allow some assistance to the non-NPT states. For example, Russia used the grandfather clause to continue with a project that had begun prior to 1992. This project involved Russia's building commercial reactors at the Kudankulam site in India. Until the May 1998 Indian nuclear tests, the United States was exploring nuclear safety assistance to India, especially involving the Tarapur power reactor that the United States had provided prior to the NSG's formation. In 2008, the NSG came full circle when it decided to carve out an exemption for India based on a request from the Bush administration. The argument from the administration was that India, whose 1974 nuclear explosive test had prompted the creation of the NSG, had been punished enough for more than thirty years and that India had growing needs for nuclear energy that India alone could not meet. Critics of this deal argued that selling outside nuclear fuel to India would free up scarce indigenous uranium supplies that India could use to make plutonium for bombs, that relaxing the guidelines (that is, rewarding India) would send the wrong message to potential proliferators such as Iran, and that leaving many Indian reactors outside of safeguards would give India the potential to produce large quantities of weapons-usable plutonium.

Has commercial nuclear power ever been used to make nuclear weapons?

This question has been controversial. Nuclear industry officials typically say no, while nonproliferation experts usually say yes. Industry officials tend to be right in the sense that, so far, no country has specifically used fissile material from a commercial nuclear power program to make weapons. Here, "commercial power" is power meant to generate electricity for homes and businesses. But nonproliferation analysts are correct to point out that certain states have used research reactors to make weapons-usable plutonium, and the expertise in nuclear-weapons programs has crossed over to power programs, and vice versa. For example, as previously mentioned, India misused a research reactor to produce weapons-usable plutonium. North Korea built a research reactor that it said was meant for electricity generation but was never used for electricity generation. This reactor has been the source of North Korea's weapons-usable plutonium. The United States weapons program provided expertise that was deployed to build reactors for warships, and these were in turn influential in shaping the U.S. commercial nuclear program. In a further blurring of the line between commercial and military nuclear activities, the United States decided in 1998 to use a commercial reactor to make tritium, a form of heavy hydrogen, for nuclear weapons. Tritium is needed to boost the explosive power of nuclear weapons. When making that decision, Secretary of Energy Bill Richardson noted that the charter for the Tennessee Valley Authority, where the reactor resides, specifies that this authority can serve defense missions. And Russia had used reactors that produced weapons-grade plutonium for residential energy purposes such as providing heating for cold Siberian cities. The production of weapons-grade plutonium in these reactors ceased in April 2010.

Presently, Iran poses the greatest concern in violating commitments to not use commercial nuclear power for weapons purposes. If it uses the cover of a power program to make fissile material for weapons, then both industry officials and nonproliferation experts would have to answer yes to the above question. Because Iran has, as of early 2011, remained inside the nonproliferation system and has consistently stated that it is only pursuing a peaceful program, Iranian acquisition of nuclear weapons would have far-reaching consequences, charting a path for other states to follow into a potentially more dangerous world.

What can be done to prevent proliferation?

Many efforts are required to prevent proliferation. While much attention in stopping proliferation focuses on limiting the spread of weapons-usable technologies and applying safeguards to peaceful nuclear programs, the real crux of the proliferation problem is political will. Importantly, states need to make a serious commitment to establishing nonproliferation as an international norm. The Non-Proliferation Treaty was a step in this direction. But as discussed, a few states remain outside this treaty. If festering security concerns are not addressed, more states may decide to pursue proliferation. One way to provide for more security is for nuclear-armed states to offer allies protection from nuclear attack. However, doing so may signal to other states the value in possessing nuclear weapons. A complementary security assurance is called a "negative security assurance" when a nuclear-armed state pledges to not attack a nonnuclear-armed state as long as that state is not allied with a nuclear-armed state. Such declaratory policies could help lower the salience of nuclear weapons. Further reducing the importance of nuclear weapons would involve nuclear-armed states' pursuing nuclear disarmament. Disarmament may not be sufficient in a world in which many states have nuclear power infrastructures. Consequently, many nonproliferation experts have warned against efforts to further spread peaceful nuclear power. In fact, the United States has a law called the 1978 Nuclear Nonproliferation Act that specifies, in its Title V, that states should assess nonnuclear energy options in comparison to nuclear energy and weigh the benefits and risks of all energy sources. The United States has never fully implemented this law. Implementation would entail the United States' performing the assessments called for in the law.

Another approach is to acknowledge that more states will pursue nuclear energy and will, thus, need access to reliable fuel supplies. To try to discourage these states from building their own fuel-making facilities, several of the major supplier states and some of their allies have proposed offering fuel assurances. Client-states would agree to refrain from enriching uranium or reprocessing plutonium in exchange for guaranteed fuel supplies. This agreement would save new nuclear entrant states from having to build their own fuel facilities and would further limit the spread of dual-use technologies. But these assurance proposals have not met with universal acceptance. Several developing states worry that their sovereign rights might be infringed. States may still pursue fuel making because of national pride and prestige, have the view that the upfront capital cost will eventually pay off, believe that this activity enhances energy security, and desire to have a latent nuclear-weapons capability. Iran is pursuing uranium enrichment for all these reasons. Although Brazil renounced a nuclear-weapons program decades ago, it has continued developing an enrichment program for the other reasons mentioned above. By comparison, the United Arab Emirates, a close U.S. ally, agreed to refrain from fuel making as part of a 2009 U.S.–UAE nuclear deal. The United States would like this deal to be a model for other states, but as indicated by the reasons for pursuing indigenous-fuel activities, this model will not be adopted by all states.

Another option is to encourage development of regional or multinational fuel-cycle facilities in which ownership, and even operational control, is shared among a group of countries. Although this option is not foolproof, it would offer the advantage of having governments look over each other's work, thereby potentially reducing the risk of one government's misusing the technology for weapons purposes. Such facilities could further enhance nonproliferation by shrouding sensitive information about how a centrifuge machine works.

Can the nuclear fuel cycle be made more proliferation resistant?

A 2004 American Physical Society report on proliferation resistance underscored the fact that there is no proliferation-proof nuclear technology, but the report emphasized that more can and should be done to improve proliferation resistance. According to Princeton University nonproliferation expert Harold Feiveson, proliferation resistance "refers to the adoption of reactor and fuel cycle concepts that would make more difficult, time-consuming, and transparent the diversion by states or sub-national groups of civilian nuclear fuel cycles to weapons purposes." Proliferation-resistant methods are classified as either intrinsic or extrinsic. Intrinsic methods

are technical and material barriers, including isotopic composition of fissile material; chemical form of material; radiation
hazards; the amount of material kept in facilities; the ability to
detect the material; facility accessibility; facility personnel's
skills, expertise, and knowledge; and the time needed to convert materials to weapons-usable form. Extrinsic methods are
institutional barriers, including safeguards, access control
and security, and location of facilities.

Can terrorists make nuclear weapons?

Fortunately, most terrorist groups are not motivated to launch
nuclear attacks. Because such an attack is an extreme form of
violence, terrorists would have to be willing to kill massive
numbers of people. Despite the popular conception of all terrorists as ruthless and irrational killers, most would not want
to inflict massive damage. As terrorism expert Brian Michael
Jenkins observed in the 1970s, "terrorists want a lot of people
watching, not a lot of people dead." Terrorism, he observed,
was often theater, in the sense that the terrorists wanted an
unusual event to draw attention to their cause and influence
the public in order to convince politicians to change policies
to the terrorists' liking. A lot of dead people would tend to
adversely affect people's views of the terrorists. This is especially the case with constituent groups that may be sympathetic with the terrorists' cause. For example, while many
people in the Basque region of Spain have been sympathetic
to the cause of forming an independent state as called for by
ETA, which had conducted numerous attacks using conventional weapons over five decades, they would not condone
massive destruction. As a sign of growing futility in using
violent means to achieve independence, Arnaldo Otegi, the

leader of ETA's political wing, announced from jail that ETA might renounce violence depending on the results of a negotiated settlement with the Spanish government, as reported by the *Wall Street Journal* on December 28, 2010.

Up until the 1980s, there were hardly any terrorist attacks that killed more than a dozen people. According to many terrorism experts, the 1979 Islamic Revolution, in which the Shah of Iran was overthrown and a theocracy was established in Iran, marked a watershed. Soon after that event, Islamic extremists used massive truck bombs in Lebanon, for example. A few other groups inclined to use massive destruction arose in the subsequent decades. Al Qaeda, which grew out of the Mujahedeen resistance to the Soviet occupation of Afghanistan in the 1980s, has shown a propensity for such attacks—it was responsible for the 9/11 attacks. More relevant for nuclear terrorism, Osama bin Laden, the leader of al Qaeda, has said that it is a duty for his group to acquire weapons of mass destruction, including nuclear weapons. The other type of group that stands out is an apocalyptic cult. Aum Shinrikyo, for example, was such a cult that was highly active in Japan until the mid-1990s. Shoko Asahara, the leader of Aum, had envisioned cleansing the world of evil by starting a nuclear war between Japan and the United States. His followers tried to buy nuclear weapons from Russia without success, and bought land in Australia that contained uranium. But Aum was fortunately far from successful in acquiring nuclear weapons. In sum, the two types of terrorists that appear most motivated to consider nuclear attacks are al Qaeda or al Qaeda-affiliated or inspired groups and apocalyptic cults.

Given motivated terrorists, following through on a nuclear attack is still hard to do. Several hurdles could trip them up.

They would either have to acquire an intact nuclear weapon from an arsenal or acquire enough fissile material to make an improvised nuclear device (IND). For the first option, the terrorists would have to breach a state's security and obtain the weapon and detonation codes or be given the weapons and this information. For the second option, they would have to penetrate security and have the technical skills needed to make the IND. The IND could use one of two designs: a gun type or an implosion type.

The gun-type device is the simplest design in that, like a gun, it fires one subcritical slug of highly enriched uranium into another subcritical piece to form a supercritical mass. The Hiroshima bomb used this design concept and was never fully tested before being detonated in war. This design can use only highly enriched uranium because the other fissile material plutonium is too reactive and would result in premature detonation. Even though this design sounds simple, tricks of the trade known by weapons experts could stymie inexperienced terrorists. To make this bomb, the terrorist group would have to acquire 40 kilograms or more of weapons-grade uranium.

In comparison, the implosion device is much harder to make and requires squeezing plutonium or HEU into a supercritical state. If the squeezing is asymmetric, the bomb will likely be a dud or, at most, produce a smaller nuclear yield than the design yield. Numerous design failures can occur. For example, one or more of the many electronic triggers needed to initiate the squeezing can misfire or not fire. Also, the initial shape of fissile material may have imperfections. To make this bomb, the terrorist group would have to acquire 25 kilograms or more of weapons-grade uranium or 6 kilograms or more of plutonium.

Assuming the terrorist group can surmount these formidable hurdles, it would next have to be able to deliver the weapon to a target without being detected and intercepted. Because the radioactive signature of uranium is relatively weak, a gun-type bomb has a reasonable chance of escaping detection. In contrast, plutonium has a stronger radioactive signal and thus may be detected, assuming that the plutonium or the bomb containing it passes close enough to a detector. In sum, terrorism involving a nuclear explosive is highly unlikely to occur but would be devastating if it did.

What can be done to prevent nuclear terrorism?

The most important actions to prevent nuclear terrorism are to apply the strongest security measures on the weapons and weapons-usable nuclear materials and to reduce the numbers of these weapons and the materials to low levels. As discussed above, the stockpiles of weapons and weapons-usable materials are immense. Thus, it is vitally important for states to act cooperatively to develop and implement the best security practices.

Other actions include more cooperation on information sharing and law enforcement among states. States should also deny terrorists safe havens in which to build nuclear explosives. Moreover, deployment of radiation detection may increase the chances of interdicting fissile material or nuclear weapons in transit. However, because fissile material has relatively weak radiation signatures, as compared to more common commercial radioactive sources, and because fissile material can also be relatively easily shielded to reduce its radiation signatures, detectors are not the strongest means of preventing nuclear terrorism. Nonetheless, as Michael Levi points out in *On Nuclear Terrorism*, each layer of the prevention system does not have to be perfect. The more layers of prevention, the less likely terrorists will succeed. Governments need to keep in mind the costs of prevention versus the risks of a successful nuclear-terrorist attack. Such an attack could soar into the trillions of dollars depending on the targeted location. In comparison, costs to secure and reduce fissile material and to deploy more effective detectors are typically no more than billions of dollars.

5

SAFETY

What is nuclear safety?

The term "nuclear safety" refers to preventing accidents at nuclear facilities, and if those accidents occur, mitigating the harm to people and the environment from any radiation release. Safety at reactors involves many activities: ensuring operators receive high-quality training, instilling a safety culture in the work habits of all personnel, performing preventive maintenance on equipment, installing layers of safety systems, retrofitting existing reactors with the best available safety systems, and designing future reactors so that they can achieve higher standards. In sum, effective safety integrates human operations and hardware performance.

How safe is safe enough, and what is safety culture?

The nuclear industry lives by the aphorism, "A nuclear accident anywhere is a nuclear accident everywhere." All utilities with nuclear power plants are in the same boat. Utility owners and industry vendors fear that this boat would likely sink if a major accident were to occur almost anywhere on the globe. The worst case from the industry standpoint would be that worry about harm from accidents would cause the pub-lic to lose confidence in nuclear power, press for cancellation of all new plant orders, and clamor to phase out existing nuclear plants. As discussed below, the major accidents of Three Mile Island and Chernobyl did not devastate the indus-try, but they did sound the alarm that the industry needed to make safety culture a top priority. So, perhaps a zero-tolerance policy on major accidents is not strictly necessary, but the industry has sought to drive down the risk of acci-dents as low as possible in order not to jeopardize the future prospects of nuclear power.

"Safety culture" is a way of doing business that, at its best, puts preventing harm to the public from accidents above keeping a power plant running if there is a suspected hazard-ous situation developing. This culture requires an attitudinal unity up and down the chain of command at a power plant. Every person, no matter how seemingly low his or her job, plays an essential role in helping to ensure the safety of the plant. If someone has reason to believe that there is a safety problem, he or she must be able to bring it to the atten-tion of the authorities without fear of disciplinary action. This no-blame and no-fault attitude helps instill a team mentality for all the workers at a plant. Another important aspect of safety culture is that plant operators should not purposely place the plant near, at, or most certainly beyond its design lim-its. This is one of the main lessons learned from the Chernobyl accident, as discussed later.

What is the defense-in-depth safety concept?

"Defense-in-depth" refers to a multilayered system in which if one layer of safety measures fails, another layer is available to prevent a major accident or at least significantly mitigate the consequences. At a commercial nuclear reactor, there may be four or five layers of protection. The first layer is the fuel itself. Nuclear fuel is designed to be robust against rupture. For example, most commercial fuel is made of uranium oxide. This material is made to be resistant to the release of highly radioactive fission products. The second layer is usually the cladding material around the uranium or plutonium fuel. This cladding is often made of zirconium or some alloy that helps prevent the release of fission products.

The third layer is the reactor pressure vessel, which is typically composed of a thick layer of steel that should be resistant to cracking or embrittlement. "Embrittlement" means that the vessel becomes brittle over a long time owing to the bombardment of high-energy neutrons. This phenomenon tends to limit the life of a nuclear plant because there is only one reactor pressure vessel, and it must stay intact to avoid a major accident from the loss of coolant. Scientists and engineers are studying how to extend the life of reactor vessels. One mechanism is called "annealing," which applies heat or thermal energy to remove the brittle areas. In the United States, because of the relatively long life of the present reactor fleet, there is considerable interest in extending the life of the reactor vessel in order to find out if many of the reactors can continue to operate for up to eighty years. Most of the current fleet is slated to receive life extensions to sixty years.

The fourth layer of protection is the containment building, which is often made of reinforced, thick concrete. This structure is designed to be airtight to prevent the release of radioactive gases to the environment. Another layer of protection that some new plants may have is a second containment structure. Also worth mentioning is the emergency core-cooling system, a layer of protection that helps keep the reactor from melting down.

What are the major types of nuclear accidents?

Major accidents are characterized as either loss of coolant or criticality. A loss-of-coolant accident, or LOCA, means that the reactor core has lost or experienced a significant disruption in its coolant. If adequate cooling is not restored relatively quickly, the reactor core may overheat and then may partially or fully melt. Recall that the fission reaction produces highly radioactive fission products. These materials release a tremendous amount of heat during their radioactive decay. This decay heat can stay at very high levels, depending on the reactor's operating history—over several hours to days. Overheating caused by a lack of coolant can cause a meltdown or rupture of the fuel, releasing highly radioactive materials. If the other layers of safety protection are not effective, the fission products may enter the environment. The Three Mile Island accident, as discussed later, was a LOCA.

A criticality accident refers to the chain reaction's going out of control. This may happen in only part of the fuel in the core. Almost all modern designs prevent this type of accident from happening because of feedback mechanisms that decrease the reactivity if the chain reaction starts to become unstable. The major exception is the Chernobyl-type reactors, which have the brand acronym RBMK, which in Russian means "high-power channel reactor." These reactors had design flaws that could make the reactivity increase in an uncontrollable fashion under certain unusual plant conditions. While safety assistance from many countries has addressed several of these flaws, many Western safety experts believe that the RBMK reactors still pose worrisome safety hazards, as discussed below. Many Western governments have called for Russia to shut down its remaining eleven RBMK reactors. Lithuania shut down its two reactors of this model as a condition of accession into the European Union, although Lithuanian safety experts have pointed out that their reactors never had a major safety problem and are of the latest, and thus claimed to be the safest, of the RBMK design series.

How is nuclear safety measured?

Plant designers and safety specialists have used probabilistic risk assessment (PRA) to analyze the likelihood of accidents. A PRA involves thinking about all the different possible sequences of events that could lead to an accident and then calculating the likelihood of each sequence's occurrence. For example, a loss-of-coolant accident could happen through various pipe ruptures or equipment failures, including a break in the main coolant pipe that carries water to the core; corrosion of the reactor pressure vessel head, resulting in a rupture; the pressurizer relief valve sticking open and draining coolant; and the feed water pumps to the steam generator breaking down and leading to drainage of the generator. For all these and other LOCA scenarios, the probabilistic risk assessment would lay out a fault tree, which shows the steps involved in each accident scenario. Analysts assign each step in a pathway a probability of occurrence. The cumulative probability of each accident scenario is determined by multiplying together the individual probabilities of the steps, assuming each step is independent of the others. In the case of a step's probability being dependent, the mathematics is more complicated. Nonetheless, the calculation can be performed. These interdependencies among probabilities arise because certain pieces of equipment—say, a major pump or a valve—affect multiple accident pathways.

It must be emphasized that probabilistic risk assessment is not easy and not without controversy. One major complicating factor is that the database on accidents is too sparse to be able to assign real-world values to the probabilities. Of course, this is a fortunate state of affairs because one would not wish for many accidents to have occurred. But unlike the rich database on automobile accidents, the database on nuclear accidents relies strongly on computer simulations and educated estimates. However, fault-tree analysis serves a very useful function in that, at a minimum, it assigns relative values to the likelihood of various scenarios. Thus, plant personnel can then know the weakest areas of plant safety and take corrective strengthening actions. This risk assessment may also allow comparison among different plant designs to determine, or at least indicate, the safety rankings of various designs. Future plant designs are tending to incorporate more passive and inherent safety features that do not depend on operator action to rescue a plant from disaster.

How safe are today's nuclear power plants?

Plants vary in the safety assessment depending on a plant's design, maintenance of equipment, training of operators, and commitment to a safety culture among a plant's management and staff. The reactors that pose the greatest safety concerns are the eleven Chernobyl-type reactors still operating in Russia. This reactor type and other reactors with designs dating from Soviet times have raised safety concerns in the

European Union (EU). The EU has required that some of these reactors be shut down as a condition for countries such as Bulgaria, Lithuania, and Slovakia to enter and remain in the EU. In general, reactors that incorporate multiple layers of safety systems appear to be the safest from a hardware perspective.

To quantify the safety record of today's nuclear power plants, it is useful to compare the number of major accidents at commercial plants versus the number of hours that they have been in operation. There have been three major accidents: Three Mile Island in 1979, Chernobyl in 1986, and Fukushima Daiichi in 2011. While other significant nuclear accidents have happened, these have not occurred at operating commercial nuclear power plants. The cumulative number of operational years worldwide is about 14,000 reactor-years. In comparison, the U.S. Nuclear Regulatory Commission seeks to have the probability of significant reactor-core damage to be better than one in 10,000 years of reactor operations. One way to visualize this probability is that if 10,000 reactors were operating, it is likely that one of them would have a major accident within one year. Alternatively, for the fleet of about 100 U.S. reactors, this probability translates to a single incident of major reactor-core damage every 100 years because 100 reactors times 100 years equals 10,000 years of cumulative reactor operations. U.S. utilities strive to keep the probability even better, at one in 100,000 years of operations. It is estimated that the best currently operating plants have a core-damage probability of about one in 1 million years of operation.

Despite improvements in safety since the Three Mile Island accident, nuclear safety in the United States has not been perfect. For instance, in March 2002, inspectors discovered that boric acid had come dangerously close to breaching the reactor pressure vessel head at Davis Besse, a nuclear power plant in Ohio. Several repairs and upgrades were needed on this plant. After about $600 million of such expenditures, the plant was ready to be reopened two years later. In January 2006, the owner of Davis Besse acknowledged several safety violations, many of which were related to the issue of the reactor pressure vessel head. Also, this plant had experienced other troubling incidents since 1977, including a faulty feedwater valve, a stuck-open pressurizer valve, failed feedwater pumps, and a tornado hitting the plant. In the latter incident, the plant's emergency diesel generators were able to supply electrical power until assured external power sources became available. The backup diesel generators illustrate the importance of defense-in-depth safety. Davis Besse is not the only plant to have committed safety violations, but it has one of the worst records.

Should a country choose one plant design instead of multiple designs, and what are the implications of such a choice?

In the ongoing debate between the U.S. and French nuclear-power philosophies, one often hears the caricature that America has one type of cheese and one hundred types of reactors while France has one hundred types of cheeses and one type of reactor. Although Americans justifiably should lament the paucity of quality cheeses in their homeland, the reality of the reactor designs is not quite so simple or stark. There are trade-offs in having many designs versus having only one or a few. Like culinary cultures, nuclear engineering in the two countries has experienced cross-fertilization and has evolved differently owing to differences in national histories.

The U.S. nuclear industry had at least a decade's head start on the French industry, which derived its commercial reactors from the U.S. technology. In the 1950s, the United States was developing two fundamentally different light-water reactor designs: the pressurized-water reactor and the boiling-water reactor. These design choices grew out of the U.S. investment in a nuclear navy, which was pursuing similar reactors. But because of several changes in regulations, evolutions in power capacities in plants, and the nascent character of this industry, many variations on these two basic designs were developed. Sometimes even two reactors of the same fundamental type at the same plant had differing designs because they were built at different times and were affected by newer construction practices and regulations. These changes usually drove up construction costs and often increased training costs for plant operators because they were not using standardized designs.

The French learned the lesson that standardization held out the promise of lower costs. Although they did not fully reap these economic savings, having only three major designs has helped them reduce the costs of preventive maintenance and training. But a design flaw could result in dozens of plants of the same design having to be taken offline and corrected. If that were to happen, France could have a substantial portion of its electricity supply shut down for an extended period of time because almost 80 percent of France's electricity is generated by nuclear power. According to the trade journal *Nucleonics Week*, generic faults have affected many French reactors at once. In particular, a reactor pressure vessel head had a cracking problem in the early 1990s. As a result, French safety authorities had to investigate many reactors to make sure the problem was not widespread. Also, concerns in 1998 and 1999 about the integrity of containment structures affected several of the 1,300-MWe series of reactors. Further, in 2001, nuclear fuel problems in the 1,300-MWe design reduced its power rating by one percentage point.

What is the China syndrome?

In 1979, a movie titled *The China Syndrome* portrayed a worst-case nuclear accident. A real-life, dreaded scenario would involve a reactor's suffering a major loss-of-coolant accident and failed emergency core-cooling system, resulting in a fuel meltdown. Under the worst imaginable situation, the reactor's containment structure would not hold the melted fuel, which would then leak into the outside environment. In the fictional nightmare scenario, the meltdown is of such severity at a U.S. nuclear power plant that it leads to the very hot radioactive material boring through the earth and emerging on the opposite side, in China. Ironically, the film was first shown to the public on March 16, 1979, just two weeks before the Three Mile Island accident. Directed by James Bridges and written by Mike Gray and T. S. Cook, the movie had an all-star cast that included Jane Fonda, Michael Douglas, and Jack Lemmon. A tag line for the movie was that, "Only a handful of people know what [the China syndrome] means . . . soon you will know." The premise is that young television reporter Kimberly Wells, played by Jane Fonda, wants to do hard-hitting news. While covering a human-interest story at the Ventana nuclear power plant, she experiences a tremor that shakes the plant. Her cameraman, played by Michael Douglas, surreptitiously films the incident. The regulatory authorities quickly issue a report saying that the plant is safe, but the cameraman suspects a cover-up. Jack Lemmon, who plays the shift supervisor on duty when the incident occurred, also believes that the plant may have suffered significant damage. The movie concludes with a dramatic standoff.

How did the Three Mile Island accident happen and what were the consequences?

In the early morning of March 29, 1979, the flow of feedwater to a steam generator was interrupted in Reactor Unit 2 at the Three Mile Island Nuclear Power Plant. Feedwater comes from condensed steam; the steam is produced in the steam generator and is used to turn the turbine that runs the electric generator. Continual flow of feedwater is needed to ensure that the reactor core does not overheat and melt down. The flow was stopped this morning because the condensate pump, which moves feedwater from the steam condenser to the polisher, turned off. The polisher removes impurities from the water before it is fed into the steam generator. Reduced flow of water through the polisher, probably because of a closed valve, caused the condensate pump to trip off. Within two minutes after loss of feedwater flow, the steam generator boiled dry. This loss of a heat sink resulted in the buildup of heat and pressure in the primary part of the plant. Primary water takes heat from the very hot reactor core. The increased pressure caused the pressure relief valve in that part of the plant to open in order to reduce the pressure. Once the pressure is reduced to a certain level, this valve should shut, but it did not, further draining water from the primary system. Compounding this problem, the reactor operators believed the valve had shut off because an indicator light showed it was closed. Thus, the operators did not have correct information. Around 4:45 A.M., about 45 minutes after the start of the accident, supervisory personnel arrived. By 6:22 A.M., the pressure relief valve was closed, but the emergency was not over. At 7:00 A.M., a site emergency was announced.

Very small releases of radiation occurred. This caused some concern about the possibility of a large release. Officials recommended evacuation of pregnant women and preschool children who lived near the plant. One of the biggest concerns surfaced on March 31, and involved the potential for a hydrogen explosion inside the pressure vessel. When steam reacts with the zircaloy cladding of the fuel at high temperature, hydrogen is formed. If a large enough amount of hydrogen reacts with oxygen, an explosion would happen. Oxygen is generated by the breakdown of water by radiation. But because there was not enough accumulation of oxygen, the likelihood of an explosion was nonexistent. Determining the amount of oxygen was a contentious issue. Nuclear Regulatory Commission experts were able to show that the presence of hydrogen in the water would be sufficient to cause controlled recombination of the oxygen within the water and thus prevent an explosion.

The immediate consequences of this event were largely financial, with the loss of revenue from the reactor. In addition, the disposal costs of the partially melted reactor core were substantial. In the longer term, the Three Mile Island accident added to the worsening financial climate for projects involving nuclear power in the United States. Although many people believe that this accident was the primary cause of the downturn in the industry, the turning point had happened earlier, as a result of a number of factors.

First, the projections for increased demand in electric power did not pan out because the United States had increased its energy efficiency following the 1973 Arab oil embargo and had experienced an economic malaise throughout much of the 1970s. Second, the regulatory environment was changing and adding costs to these nuclear projects. Third, the long lead time for construction of nuclear plants had slowed interest in them. Notably, no deaths are directly attributable to this accident. The containment building around the reactor maintained its integrity and prevented large releases of radioactive materials into the environment. A later question addresses how the nuclear industry responded to this accident.

How did the Chernobyl accident happen and what were the consequences?

In the early-morning hours of April 26, 1986, the worst nuclear accident the world has experienced began to unfold. At that time, four reactors were operating at the Chernobyl Nuclear Power Plant, located in Ukraine, which was part of the Soviet Union. Reactor Unit 4 was undergoing tests that placed the reactor in an unusual condition. Starting the previous night, the operators had begun testing a safety feature of the reactor. They wanted to determine if the electric generator starting in a low power condition and coasting down to lower power would provide enough electricity to run the coolant pumps. The reactor would normally have backup power supplies from off-site electricity and emergency diesel generators. The plan for the test was to not rely on off-site power and to factor in the several seconds' delay of start-up of the diesel generators. During that delay, the plant's electric generator in a coasting-down mode should be able to provide adequate power to maintain safety. But the test itself ran into delays. Many hours elapsed before the operators were able to take the electric generator down to lower power levels. Meanwhile, xenon had built up inside the reactor. Xenon, a product of fission, acts like a reactivity "poison," in that it absorbs neutrons and makes it harder to sustain a controlled chain reaction. To compensate for the xenon buildup, the operators raised the reactor's control rods, to a high level that violated operating guidelines. Control rods are made of materials that absorb neutrons; therefore, as the rods are raised, more neutrons are available to increase the reactivity of the reactor. But raising the rods to such a high level substantially reduced the capability of the plant to operate safely. Although the operators should have stopped the test at this point, they continued. A further reduction in water flow and reactor reactivity led to additional automatic raising of some control rods, placing the plant in an even greater potentially unsafe condition.

With the control rods in a precarious position, the operators reduced steam pressure to the turbines. This step resulted in a decrease of water flow through the reactor core and generation of some steam inside the core. Because of a design flaw, this formation of steam triggered an upsurge of reactivity. The design flaw is known as a "positive void coefficient." When a void—a vacuum or low pressure volume—is formed owing to replacement of liquid water with a steam bubble, the neutrons that drive reactivity will not slow down as quickly as they normally would in water. Faster neutrons should reduce reactivity because they have a lower probability to cause fission than slow neutrons. But the Chernobyl, or RBMK, reactors were designed so that graphite was the main moderator of neutrons and water was mainly used as a coolant. Water also has the property of capturing a small fraction of the neutrons. So, when the steam bubbles formed, the reactor had more neutrons available for fission because of the lack of water to capture them. And these neutrons were moderated and slowed down by the graphite. Thus, the reactivity surged. This power upswing alone was not sufficient to cause the catastrophe, however. To stop the surge, operators inserted the control rods. But because of a design flaw in the rods, their insertion from the high position led to a further huge increase in reactivity. The ends of the rods were made of graphite which, when inserted into the core, moderated more neutrons and thus raised reactivity.

In less than one minute, two large explosions occurred. One was a steam explosion, exposing the reactor fuel to air. The other reaction resulted from liberated hydrogen gas, a flammable substance, reacting with oxygen. The explosions blew the top off the reactor building, and the burning debris ignited fires on the remnants of the reactor's roof. Teams of heroic firefighters eventually put out the fires and prevented the fire from spreading to the other reactor units. Thirty-one firefighters and emergency responders lost their lives because of exposure to high doses of ionizing radiation.

In addition, because many children did not receive potassium iodide tablets to prevent exposure to radioactive iodine from the contamination, about 1,800 thyroid cancers developed in excess of what would be normally expected in the exposed population. These cancers could have been prevented, but they happened because of the Soviet secrecy in not informing people before it was too late. Experts have disputed whether there were other directly attributable health effects; estimates range from very few additional incidents of cancer to more than 20,000. The problem in determining the total health effects is in picking out the health incidents that resulted from Chernobyl versus all other health hazards experienced by the population. Mortality rates in that region have soared, largely because of increased stress, alcoholism, and poor nutrition.

Massive radioactive contamination covered the surrounding regions near the reactor site. About 135,000 people living within thirty kilometers of the reactor were evacuated. The most heavily contaminated area was declared an exclusion zone. Ukraine and Belarus bore the brunt of the contamination, but many other European countries had detectable amounts of radioactive fallout. The total cost estimates for the accident are upwards of a few hundred billion dollars.

Mikhail Gorbachev, the Soviet leader during that time, later said that the Chernobyl accident was a tipping point that helped him in his efforts to break down pervasive and oppressive secrecy in the Soviet political system and to open up Soviet society. This opening up helped spur the collapse of the Soviet Union within a few years after the nuclear accident. So, in a sense, one of the greatest unforeseen, indirect consequences of this accident was the downfall of the Soviet system.

The accident also contributed to the slowdown in the use of nuclear power, especially in some European countries such as Austria, Germany, and Sweden, although only Austria completely outlawed the use of nuclear power. Ironically, Austria today receives part of its electricity via the inter-country electrical grid from the Eastern and Central European nuclear-power countries. But Austrian politicians dare not admit this fact because the majority of the population is opposed to nuclear power and would not vote for politicians who support it. The denial runs so deep that electrical utility bills in parts of Austria—especially in parts that are very anti-nuclear—lists the amount of electricity generated from nuclear power as zero or close to zero.

Could the radioactive contamination have been kept within the reactor building? The explosions were extremely powerful and may have ruptured even strong containment structures. But unfortunately, another design flaw was that the RBMKs did not use containment structures, in contrast to almost all Western-designed nuclear plants. Containments, as mentioned earlier, are the last line of defense for preventing release of radioactive materials to the environment. In other safety hazards, RBMK plants had inadequate fire-protection systems, as well as subpar operating, training, and emergency procedures. Moreover, they had insufficient separation and redundancy of safety systems. Further, these plants had inadequate systems for monitoring key safety parameters. Perhaps the most damaging flaw was, according to the evaluation of the International Nuclear Safety Advisory Group, the "deficient safety culture, not only at the Chernobyl plant, but throughout the Soviet design, operating, and regulatory organizations."

In December 2000, the last of the operating reactors at the Chernobyl site was shut down. Ukraine agreed to do so under the terms of a memorandum of understanding with the Group of Seven (G-7) countries, which are the leading industrialized states of Canada, France, Germany, Japan, Italy, the United Kingdom, and the United States. Under this agreement, the G-7 countries have provided assistance in mitigating the risks at Chernobyl, offered financial support for energy-efficiency projects, and helped address the social and economic effects resulting from Chernobyl's closure. Assistance was also provided to finance replacement nuclear reactors of much safer designs and to shore up the unstable shelter, or "sarcophagus," built after the accident and around the damaged reactor.

What happened to Soviet-designed reactors after the Chernobyl accident?

The Chernobyl accident attracted significant international political attention and safety assistance. The Group of Seven industrialized countries meeting in Lisbon in 1992 was the first major high-level international meeting to focus on the safety of Soviet-designed reactors. One of the results was to set up the Nuclear Safety Account in 1993, as a multilateral source of funding for safety improvements to these reactors. The European Bank for Reconstruction and Development, in particular, provided the financial mechanism to direct much of the international money for safety assistance. In addition, G-7 nations, as well as others, contributed funding through bilateral agreements with countries possessing Soviet-era reactors. Several hundred millions of dollars worth of assistance were provided.

International aid helped pay for a number of safety improvements in the RBMKs and other Soviet-designed reactors. Although much attention has been given to the safety deficiencies, many of these reactors have strengths that Western-designed reactors do not have. For instance, the VVER-440 reactors, which have a significantly different design from the RBMKs, have six primary coolant loops, increasing the amount of water available to keep the reactor cool. The VVERs have isolation valves that permit removing one or more loops from service while allowing the plant to continue operating. Only a few Western plants use this feature. The steam generators are placed horizontally, whereas they are placed vertically in Western designs. This configuration allows for improved heat transfer.

Safety deficiencies in the VVER-440 series (especially the earlier generation model 230 reactors) included the lack of a containment structure, emergency core-cooling systems that did not meet Western standards, increased risk of embrittlement of the reactor pressure vessel, plant instrumentation and control systems below Western standards, and inadequate fire-protection systems. Throughout the 1990s and into the 2000s, these deficiencies raised concerns in Western Europe, especially as the former Warsaw Pact countries sought to enter the European Union. The EU made entry conditional on either shutting down certain reactors or making major safety improvements. Experts from the EU and the United States provided safety assistance that dramatically improved the safety and performance of many of the Soviet-designed reactors. But two designs were deemed unacceptable; these were the RBMK and the VVER-440 model 230.

Two RBMKs, or so-called Chernobyl-type reactors, were still operating in Lithuania. The Lithuanian government agreed to shut down these reactors as a condition of entry into the EU, and the last Lithuanian RBMK was shut down in December 2009. Bulgaria and Slovakia were the only two new EU entrant states with VVER-440 model 230 reactors. Bulgaria shut down Kozloduy Units 1 and 2 in December 2002. Despite many safety enhancements to Units 3 and 4, Bulgaria reluctantly shut them down in December 2006, as well, and joined the EU in January 2007. But under the terms of the EU accession agreement, those two units could return to operation if they are needed to deal with a national energy crisis. Like Bulgaria, Slovakia had, with outside assistance, made many improvements to its VVER-440 model 230 reactors, but its government reluctantly followed through with closure. Unit 1 at Bohunice was shut down in December 2006 and Unit 2 in December 2008. The prime minister at the time of the first shutdown called the action "energy treason" and blamed the previous government for making a bad decision.

The VVER-1000 reactors, designed in the Soviet Union, were similar to Western pressurized-water reactors, but they did not fully meet Western standards, mainly because of deficiencies in plant instrumentation and controls, fire protection, and operating procedures. This series did, however, employ a containment building and had built on the strengths of the earlier generation VVER-440s. In recent years, Russia has been building domestically and exporting internationally the VVER-1000. Russian engineers in partnership with counterparts in other countries have made safety upgrades to this design.

How did the nuclear industry form self-policing organizations?

Soon after the Three Mile Island accident, the owners of U.S. nuclear power plants responded by forming the Institute for Nuclear Power Operations (INPO) in December 1979. Headquartered in Atlanta, Georgia, INPO's founding mission is to achieve excellence in power-plant operations and focus on safety as a top priority. Its leadership mostly came from the U.S. nuclear navy, which instills safety as a primary responsibility for everyone involved in maintaining and operating naval nuclear plants. The INPO conducts numerous commercial nuclear-plant evaluations and ranks plant safety performance based on objective criteria. The goal is to help plant operators improve their performance. The INPO assesses plant personnel from the most junior operators to chief executive officers. While letting the industry inspect itself without oversight is not acceptable, the INPO has complemented the vital role of the U.S. Nuclear Regulatory Commission.

Similarly, worldwide it is essential that all countries with nuclear power plants have strong national regulatory authorities. However, the INPO has established a precedent for plant owners around the globe to take a more active role in ensuring excellence in safety. To help prevent another major accident such as Chernobyl, plant owners responded by forming the

World Association of Nuclear Operators (WANO). Patterned after INPO, WANO seeks to instill high standards of safety and performance in nuclear plants around the world. WANO has conducted hundreds of inspections at almost all commercial nuclear plants in about thirty countries. WANO has regional offices in Atlanta, London, Moscow, Paris, and Tokyo. In early 2010, WANO was in the process of changing its governance structure in recognition of the increasing role of multinational ownership of major nuclear companies and utilities.

Can the nuclear industry survive if another major accident occurs?

Nuclear industry leaders generally have a zero-tolerance policy for major accidents. They believe that another accident of the severity of Three Mile Island or Chernobyl would likely doom the prospects for more growth. The unfolding consequences of the Fukushima Daiichi accident will test this belief. It is interesting to compare this attitude to air travel. People have come to accept that occasionally airplane accidents will occur, resulting in hundreds of people dying. But in comparison to the millions of people traveling by air and the tens of thousands of flights daily, a few accidents annually seem acceptable. So, why do many people feel that a major nuclear accident would be intolerable? A fear of massive amounts of ionizing radiation—an invisible substance that may cause serious health effects—is one reason. A related concern is that large land areas may become uninhabitable for decades. Perhaps a major difference between air travel and nuclear-generated electricity is that people do not have a readily available alternative for moving rapidly from one place to another distant place, but they do have readily available alternatives for generating electricity.

It is also useful to compare risk perceptions among car, air-plane, and nuclear accidents. Car accidents are far more fre-quent than airplane accidents. But people tend to feel in control when they drive cars, and they feel less in control when they are passengers on a plane. In contrast, many people near a nuclear power plant often feel that they have little or no choice about whether to live near the plant and no control whatever over what is being done to maintain the safety of the plant.

How long can nuclear power plants operate?

The nuclear age is still relatively young. Commercial nuclear power only began in the 1950s, and the cumulative operating experience for the world's reactors is about 14,000 reactor-years. While this amount of time may seem long, it is short compared to estimates for the amount of time in which a major accident may be expected. As mentioned earlier, the major core damage probability for almost all operating reactors is at least one in 10,000 reactor-years or better. Thus, every 10,000 reactor-years of operation, it is likely that there will be an accident that causes core damage. Putting aside the Chernobyl accident as an aberration because of its many design flaws and the poor safety culture among Soviet reactor operators, only two major accidents—Three Mile Island and Fukushima Daiichi—have occurred at a modern plant. With continuing safety improve-ments, the industry is trying to avoid any additional major acci-dents. But nuclear power plants are complex and are operated by imperfect, if well-trained, humans. So, regulatory authori-ties are cognizant that every plant has a life span, and it is prudent not to push up against the end of life, when problems are more apt to develop. The biggest worry is that the reactor pressure vessel could become brittle after many decades of bombardment by neutrons. The neutrons dislocate the atoms in the metallic vessel. Over a long time, these dislocations become sites where cracks can appear. If the vessel breaks, the reactor would experience a loss-of-coolant accident and could melt down if emergency cooling systems did not function properly.

When the United States began licensing nuclear plants in the 1950s, the view was that plants could operate for at least forty years. As the due date approached, owners of the plants could request a life extension from the regulatory authority. Because of the more than thirty-year hiatus in plant orders, though, the U.S. reactor fleet is approaching its nominal life span. In recent years, more than fifty reactors in the United States have asked for life extensions for another twenty years of operation. Almost all have received the extension. Nonetheless, by the early 2030s, the United States will have to start decommissioning many reactors if these reactors do not receive additional life extensions. Some regulatory authorities such as former NRC Chairmen Nils Diaz and Dale Klein have talked about determining whether most U.S. reactors can operate for a total of eighty years. More research and development is needed to figure out whether and how to accomplish this life extension. In contrast to the U.S. licensing of forty years, the French regulatory authority issues ten-year licenses and performs checks near the end of this period to decide whether to issue another ten-year license. However, the French reactors still face the same end-of-life considerations.

How can future nuclear power plants be made safer?

To ensure the continued viability of the nuclear industry, designers have sought to make future plants safer. The ongoing debate is whether to emphasize passive or active safety components and systems. "Passive safety" refers to safety components that do not require active operator intervention to make them function. These systems rely on natural forces such as gravity and heat convection to enhance a plant's safety. It is important to underscore that "passively safe" does not necessarily equate to "inherently safe," although some designs are striving for such an achievement. In an inherently safe plant, the operators could just walk away and not have to intervene even after many days or weeks. Two new passively safe designs are claimed to not require operator intervention for up to three days after loss of coolant. These designs are General Electric's Economic Simplified Boiling Water Reactor (ESBWR) and Westinghouse's AP-1000, where the AP stands for "advanced passive." The ESBWR can achieve greater natural-coolant circulation by using a taller reactor vessel and a shorter reactor core, and by reducing water-flow restrictions. In addition, this design employs an isolation condenser system that would control high-pressure water levels and would enhance radioactive decay heat removal. Moreover, the ESBWR has a gravity-driven cooling system to provide low-pressure water-level control, according to the U.S. Nuclear Regulatory Commission. The AP-1000 uses similar concepts and makes use of natural driving forces so that actively operated pumps and diesels are not required. However, all the new reactor designs still hold to the principle of defense-in-depth safety and thus have redundant safety components.

In contrast to the ESBWR and the AP-1000, the French-designed Evolutionary Power Reactor relies on active safety features. This design incorporates multiple engineered safety features, including a double-walled containment and a "core catcher" for trapping, holding, and cooling radioactive materials from the reactor core in the event of a severe accident resulting in reactor vessel failure. In sum, an active safety system requires reliable electrical power to operate pumps, valves, and other safety components. Nonetheless, all plants use passive safety components in that the reactor fuel design and the containment are built such that they do not require active electrical and mechanical control systems to function. Regarding designs claimed to be inherently safe, the Pebble Bed Modular Reactor (PBMR) uses helium gas to transfer heat from the reactor core. The fuel consists of balls of graphite and ceramic-coated uranium particles. The fuel design is inherently safe in that, if it begins to overheat, the probability of neutron capture by uranium-238 atoms increases and thus the rate of fission of uranium-235 atoms decreases. Consequently, this feedback reduces the reactivity and the heat production, driving the reactor into a stable condition. The geometry and contents of the fuel pebbles provide additional inherent safety. While the designers are so confident of the PBMR's safety that they believe that a containment structure is not necessary, the public may not be so convinced and may demand this safety feature. Building a containment structure adds significantly to the cost of a nuclear plant.

Can nuclear power expand too fast to keep plants safe?

As of early 2010, more than two dozen countries have expressed interest in acquiring their first nuclear power plants. While a few of these countries, such as Indonesia, the Philippines, and Turkey, had ventured down this path previously and thus have built up some indigenous expertise, most of the potential new entrants are starting practically from square one in training skilled people who can safely operate and inspect plants. It is essential that these governments have independent regulatory agencies with sufficient authority to order a plant shut down if there is a safety problem that may jeopardize the plant and may harm the public.

The more immediate safety concern has centered on China. China has far from a stellar safety record, as illustrated by lead contamination in children's toys, melamine-tainted milk, and accidents at coal mines. The Chinese government has planned for a massive ramp-up of its nuclear electricity capacity. In 2009, Beijing announced targets of 70 gigawatts of capacity by 2020 and 400 gigawatts by 2050. In comparison, China has, in late 2009, about 9 gigawatts of nuclear power from eleven reactors and has another ten reactors under construction. In October 2009, to meet this increased demand, Prime Minister Wen Jiabao ordered a quintupling of the safety personnel. This amount may still not be enough, according to independent experts. It can take several years to train safety inspectors. This breakneck pace is further worrying in light of scandals in the Chinese nuclear industry. In August 2009, for instance, Chinese authorities fired Kang Rixin, the president of the China National Nuclear Corporation. He was allegedly involved in bid-rigging of power-plant construction. According to the *New York Times*, Mr. Kang's alleged actions were not known to have created hazardous conditions at the plants. The rapid expansion of nuclear power in China, and the potential for newer nuclear-power states, underscore the need for increased, urgent international cooperation in nuclear safety, with the recognition that it is each individual government's responsibility to enact, implement, and enforce the highest safety standards.

How can nuclear facilities be made resistant to natural disasters such as earthquakes and tsunamis?

Understanding the potential natural hazards near a proposed or existing nuclear facility is the first and most important step in protecting both the facility and the public. For instance, a nuclear power plant should generally not be constructed near a major earthquake fault. Nuclear facilities can be protected against natural disasters by using quality construction and effective safety systems. For example, to protect against earthquake damage, the installations need to be able to vibrate without damaging vital safety systems, such as the reactor pressure vessel and the containment structure. To protect against tsunamis, nuclear plants along coastlines need seawalls tall enough to block very high waves. They also need emergency diesel generators located where they are not vulnerable to flooding. Finally, the plant's emergency core cooling systems must remain functional even under severe natural disasters. While engineers work to increase the safety of nuclear facilities, it is probably impossible to prepare for all possible contingencies. Nature is always capable of inflicting unprecedented damage, for which we are not fully prepared.

How did the Fukushima Daiichi accident happen?

On March 11, 2011, a massive earthquake occurred at 2:46 P.M. local time off the northeastern coast of Honshu, Japan's largest and most populated island. This earthquake measured approximately 9.0 on the Richter scale, the largest ever recorded in Japan's 140-year history of monitoring seismic events. The sudden and large movement of the Pacific tectonic plate also subsequently triggered a gigantic tsunami. This tall and fast-moving wall of water slammed into the land at and around Sendai, a city in northeastern Honshu. Two weeks later, more than 10,000 people had been confirmed dead and about 17,000 were still missing.

The intense shaking from the earthquake had surpassed the safety thresholds of seismometers located at Japan's Fukushima Daiichi, Fukushima Daini, Onogawa, and Tokai nuclear power plants. Consequently, these seismometers immediately sent signals to the eleven operating reactors at these plants to shut down. Three of Fukushima Daiichi's six reactors were operating when the earthquake struck. Reactor units 1, 2, and 3 were the oldest at that site, having begun operations in the 1970s. The shutdown procedure proceeded correctly. Nonetheless, a crisis unfolded when the tsunami flooded the emergency diesel generators at the Fukushima Daiichi plant. As a result of the earthquake, off-site electrical power, which was essential in order to provide electricity to run coolant pumps at the plant, was also unavailable.

As mentioned earlier in this book, several days to a few weeks after a reactor shuts down—depending on the operating history—the reactor still generates a large amount of heat from the radioactive decay of fission products. Immediately after a shutdown, this decay heat is about 6 percent of the heat produced just before the shutdown. For example, reactor unit 1 at Fukushima Daiichi was generating about 1,200 megawatts of thermal power. So, right after shutdown, the reactor core was still generating about 72 megawatts of thermal power. This tremendous amount of heat must be removed from the core or the fuel rods could rupture and eventually meltdown if cooling remains unavailable. Despite the lack of off-site electrical power and the flooded diesel generators, the plant had batteries with about eight hours of stored electrical energy. But this backup power was depleted before other means of electrical power could be restored.

With no electrical power to run the cooling pumps, these reactors' cores began to overheat. Much of the water surrounding the cores turned to steam. The steam then interacted chemically with the zircalloy cladding that is designed to protect the fission products in the nuclear fuel from escaping into the environment. This chemical reaction produced hydrogen gas, which is flammable. To reduce the buildup of steam pressure and hydrogen gas inside the reactor's pressure vessel and primary containment structure, the plant's operators vented some steam and hydrogen gas. The main danger was that the primary containment would rupture. But within a few days, the hydrogen released into the secondary containment structure ignited and blew holes in the secondary containments of reactor units 1 and 3. While the cores themselves were not exposed, that was not true for the spent fuel pools located inside the secondary containments. The concern then was that the spent fuel pools could lose cooling water; and if the spent fuel caught fire, large amounts of radioactive material could be dispersed into the environment. Reactor unit 4's spent fuel pool did experience a significant loss of coolant because of damage from the earthquake. But during the first week of the crisis, there was a difference of opinion as to whether the pool had emptied or had just suffered a major loss of coolant. On March 16, NRC Chairman Gregory Jaczko testified to Congress that he had information of a complete loss of coolant, while Japanese nuclear industry officials disagreed. This discrepancy illustrated the difficulties in obtaining reliable data.

Also, during the first week of the crisis, reactor unit 2 suffered a hydrogen explosion in the steam suppression system attached to the primary containment. This system was designed to prevent the buildup of steam by absorbing the steam's energy. But the steam suppression systems became saturated in units 1, 2, and 3, and a breach in unit 2's steam suppression system provided a potential pathway for large amounts of radioactive materials to leave the primary containment in the event of a meltdown of the fuel. Relatively smaller amounts of radioactive materials were being emitted every time the operators vented steam; in particular, radioactive iodine and cesium were being released. But during the second week of the crisis, levels of radioactive iodine in water as far away as Tokyo—about 140 miles (about 220 kilometers) from Fukushima—prompted authorities to advise parents to not give the contaminated water to babies and young children, lest significant amounts of radioactive iodine accumulate in the children's thyroid glands.

Five weeks into the crisis, plant operators managed to inject fresh water into reactors 1, 2, and 3; in the first few weeks, they had used seawater as an emergency means of cooling. The temperatures in these reactors' cores did decrease and stabilize, but were still higher than the cold shutdown level. Operators also sprayed fresh water into the spent fuel pools of units 3 and 4. On April 12, the Japanese Nuclear and Industrial Safety Agency revised its rating of the crisis to 7, the highest level on the International Nuclear Event Scale. While this rating assessment matches the numerical rating of the Chernobyl accident, the Fukushima Daiichi accident had released an estimated one tenth of the amount of radiation that was released in Chernobyl.

This nuclear accident is unprecedented, in that it involved more than one reactor simultaneously experiencing major damage. In comparison, the accidents at Three Mile Island and Chernobyl each involved only one reactor. While it is too early to know the full implications of this accident, for both Japan and the world, some lessons to be learned were apparent early on in the crisis, as discussed below.

Were there design flaws in the Fukushima Daiichi reactors and were there safety concerns prior to the accident?

These reactors are boiling-water reactors (BWR). While chapter 1 discussed the fundamental aspects of a BWR, it is important to delve into the Mark I type of BWR because five of the six Fukushima Daiichi reactors were this type, and those were the ones to experience the most damage. The Mark I was designed to save money by having a relatively small primary containment structure. As a protective measure, the designers added a steam suppression system that could absorb energy from the steam generated should there be an accident. Thus, the design would reduce the pressure buildup from the steam, allowing the designers to reduce the size of the containment. But that smaller primary containment could be too limited to handle the volume of steam in the event that the steam suppression system became saturated. This is exactly what happened at Fukushima Daiichi. Faced with this situation, operators would be forced to vent the steam from the primary containment to prevent a breach of the containment structure.

Concerns about this design date back to at least 1972, the year after reactor 1 at Fukushima Daiichi began operations. At that time, Stephen Hanauer, an Atomic Energy Commission safety official, recommended that the Mark I be discontinued.

But because this design was being used in many other reactors, industry officials opposed a recommendation that would result in loss of revenue from these reactors. Further studies throughout the 1970s and into the 1980s continued to point to potential safety problems for the Mark I under certain accident scenarios. To mitigate the potential damage from such accidents, nongovernmental scientists Frank von Hippel and Jan Beyea recommended in 1982 that the containment structures be retrofitted with filter systems to capture radioactive gases. But the utility owners again did not want to spend money on these filters, and the NRC did not require that the filters be added. As of early 2011, twenty-three U.S. reactors still use the BWR Mark I design.

In addition to concerns about the reactor design at Fukushima Daiichi, the vulnerability of the emergency diesel generators to the tsunami have caused alarm. In particular, the generators were not in raised positions. Instead, their placement left them exposed to the floodwaters. While a seawall provided some protection from flooding, it was too low to stop the onrush of a tsunami greater than 20 feet (more than 6 meters) high. Indeed, this tsunami was much higher than any other recorded in Japan's modern history. But in A.D. 869, that region of Japan had experienced a tsunami so powerful that it knocked down a castle. According to a March 24, 2011, article in the *Washington Post*, despite evidence that this region was vulnerable to both massive earthquakes and large tsunamis, executives of Tokyo Electric Power Company (TEPCO), which owns Fukushima Daiichi, had scoffed at warnings by prominent seismologist Yukinobu Okamura. These warnings came in June 2009, during investigations by Japan's Nuclear and Industrial Agency into the protections against natural hazards needed at Japanese nuclear power plants.

What are the concerns about nuclear safety culture in Japan?

The accident at Fukushima Daiichi has renewed the world's concerns that Japan's nuclear regulatory agency lacks sufficient independence and authority to call for necessary retrofits or closures that can ensure the safety of its nuclear power plants. One month before the earthquake and tsunami, Japanese regulators approved a ten-year extension of the license for reactor 1 (the oldest reactor, with about forty years of operation), despite warnings of safety problems at the plant and specifically with this reactor. In particular, according to a March 21, 2011, article in the *New York Times*, there were known stress cracks in the emergency diesel generators, potentially making them susceptible to corrosion from seawater and rainwater. Moreover, after the license extension was approved, the Tokyo Electric Power Company (TEPCO), which owns the plant, admitted that it had not inspected thirty-three pieces of equipment associated with the plant's cooling systems. Although regulators cited insufficient inspections and poor maintenance, they authorized continued use of the plant. Critics of Japan's regulatory system have warned repeatedly about the unhealthy ties between regulators and utility owners. Japan's ten regional electric utilities have monopolistic control in the regions where each utility operates, and this power gives them strong sway over local and national governments.

Attempts to cover up problems at Japan's nuclear plants have a relatively long history. In particular, dating back at least to the 1980s, TEPCO had allegedly falsified data from a number of nuclear plants, according to an investigation done in 2003. In response, then TEPCO president Tsunehisa Katsumata pledged a new code of ethics for his company, and TEPCO CEO Hiroshi Araki and four other executives resigned. But a few years later, new reports surfaced of other incidents of alleged falsifying of data. Concerns thus remain that the corporate culture has yet to improve substantially.

What are the likely implications for the nuclear industry as a result of the Fukushima Daiichi accident?

As of two weeks after the start of the accident, the full implications are far from clear or certain. However, some governments, including China, Germany, and Switzerland, have called for a moratorium on the construction of new nuclear plants until thorough safety checks have been done. Even before the accident, Germany had decided in the late 1990s to phase out its nuclear power plants. The accident further prompted the German government to immediately shut down its seven oldest reactors while safety checks are being performed. Conversely, China will likely continue to forge ahead with its ambitious plans for dozens of nuclear plants. Nonetheless, Chinese leaders felt compelled to respond to its public's concerns about radioactive contamination potentially coming from Japan.

In the United States, some politicians such as Representative Edward Markey, a Democrat from Massachusetts, and Senator Joseph Lieberman, an independent from Connecticut who was previously a Democrat, have recommended that a moratorium be placed on the construction of any new U.S. nuclear plants. Meanwhile, President Obama has expressed continued support for construction of new nuclear plants. During the second week of the Japan nuclear crisis, the Nuclear Regulatory Commission announced that it would conduct a ninety-day review of all U.S. nuclear plants.

Ultimately, this accident may lead to much improved safety standards and greater protections against natural disasters at both existing and future plants. In particular, new regulations might require utilities to remove older spent fuel from overcrowded spent fuel pools so as to reduce any risks of dispersal of radioactive materials resulting from accidents, attacks, or sabotage. (Chapter 7 discusses the potential vulnerabilities of spent fuel pools to attacks or sabotage.) The industry itself might also voluntarily implement additional safety features to bolster public confidence. One of the clear lessons of the accident is that government and industry officials need to be much more transparent about nuclear operations and what needs to be done to protect the public.

Why did the Fukushima Daiichi accident raise renewed concerns about the use of plutonium in nuclear fuel?

Reactor 3 at Fukushima Daiichi had recently been fueled with mixed oxide fuel containing a mixture of plutonium oxide and uranium oxide. If there were a breach of the primary containment and a release of radioactive materials from the reactor core, plutonium might be dispersed into the environment. (As discussed in chapter 2, Japan is one of the few nuclear power–producing countries that have been using recycled plutonium in certain nuclear plants.) Plutonium poses little health hazards if outside the human body, because the protective outer layer of skin blocks most of the emitted ionizing radiation, especially the most prevalent alpha radiation. But if it is ingested or inhaled in appreciable amounts of micrograms or more, plutonium can be extremely hazardous, causing cell damage from alpha radiation and having toxic effects on kidneys and other organs. In comparison to nuclear fuel containing only uranium oxide, mixed-oxide fuel is more hazardous owing to plutonium's being more radioactive than uranium. But the degree of hazard is debatable. Uranium-fueled reactors produce plutonium resulting from the radio-active transformation of uranium-238. Over the life of a fuel assembly initially containing uranium fuel, plutonium will, in as little as a few weeks, start to contribute significantly to the power produced in the reactor. By the end of the fuel assembly's time in the reactor, almost 1 percent of it by weight is plutonium. Thus, even uranium-fueled reactors have a risk of plutonium dispersal in the event of a severe accident. Nonetheless, reactors using mixed-oxide fuel will contain more plutonium that can be dispersed. Another mitigating factor is that plutonium tends to not be very dispersible in air or soluble in water.

Additionally, plutonium-based fuels raise a safety concern about possibly shortening the usable life of a reactor. Because fission of plutonium produces more neutrons, on average, than fission of uranium, the excess neutrons smacking into the reactor pressure vessel may damage this vital safety system. As discussed earlier, neutron bombardment of the pressure vessel can cause it to become brittle.

6
PHYSICAL SECURITY

What is nuclear security?

The meaning of "nuclear security" depends on the context. When referring to fissile material that can power nuclear bombs, this concept entails making sure that the material is secure against diversion or theft. In regard to deterring nuclear war, the term refers to ensuring that nuclear arsenals are secure. But here, the focus is on the security of nuclear facilities such as nuclear power plants, radioactive waste storage and disposal facilities, and commercial fuel-cycle facilities.

Securing a nuclear facility first requires understanding the interplay among a set of factors: potential attackers or saboteurs, vulnerabilities to that facility, and ways the attackers or saboteurs may breach defenses and exploit those vulnerabilities. Because of the complexity of the security task, several different types of professions must work cohesively. Intelligence analysts need to assess continually the threats to the facility and provide that assessment to the facility's operators and guards. Operators and guards need to coordinate their actions so that the facility remains safe and secure. On-site guards require rigorous training against realistic simulated attacks. These guards often are the first line of defense against an assault on a facility. Because the attackers could overwhelm the defenders, emergency-response forces should be available and ready to arrive on the scene within a relatively short time period. This team of intelligence analysts, facility operators, guards, and emergency-response personnel work best together when they form and act cooperatively, as organized by a design-basis threat assessment.

What is a design-basis threat assessment?

According to the International Atomic Energy Agency, the design-basis threat (DBT) assessment is founded on four concepts: understanding the potential adversaries, whether insiders or external attackers; analyzing the capabilities of those adversaries; carrying out measures needed to prevent malicious acts or to mitigate the consequences of those acts; and establishing performance requirements for physical protection of the facilities. The DBT should also have well-defined methods to grade the effectiveness of guard forces, in particular, and the overall physical protection system's performance, in general. Moreover, the DBT assigns clear responsibilities to each position within this system. For example, guard forces are taught when they are permitted to use deadly force to stop an attack. Operators are instructed in actions they should take to place the facility in a safe condition if it is under attack or subject to sabotage. Governments are responsible for ensuring timely analysis of potential threats and for conveying that assessment to operators and guard forces. The IAEA has published a detailed guide for conducting a formal DBT.

How are safety and security different?

In daily language, people often use the words *safety* and *security* interchangeably. But these terms have specific meanings in the nuclear field. "Safety" refers to making sure that a nuclear facility is unlikely to have an accident, and in the event of an accident, that there are ways to keep the consequences to low levels. Accidents are unintentional. In contrast, "security" refers to ensuring that a facility is protected against attack or sabotage, and in the event of an attack or sabotage, that there are procedures to mitigate the damage. Breaches of security result from intentional actions by adversaries.

The overlap between safety and security involves common efforts taken to ensure that a facility is both safe and secure. For instance, the defense-in-depth safety concept employed at nuclear power plants offers protection against severe consequences owing to an attack. The training of plant operators can and should emphasize actions they can take to keep the plant safe and secure.

One of the major differences between these concepts is in their accompanying cultures. A safety culture has succeeded best in work environments in which potential hazards are discussed openly and plant personnel are encouraged to bring attention to potential problems early, without fearing repercussions. In fact, the employees should be rewarded for drawing attention to safety concerns. While at times personnel have been reprimanded for doing such, whistleblower protection laws are needed to assure workers. The highest performing and safest plants tend to foster openness. By comparison, the culture surrounding security personnel has traditionally been marked by a wariness in revealing security problems; personnel do not want to reveal information that could give an advantage to an adversary. Nonetheless, some openness is needed here, too, to allow greater cooperation between guards and operators and to let the public know that efforts are being taken to keep a plant secure. Moreover, by broadcasting to adversaries that a facility is secure, enemies such as terrorist groups would likely refrain from attacking a hardened target. Thus, an essential element of defense is to dissuade opponents from launching attacks.

Why would someone attack a nuclear power plant or related nuclear facility?

The 9/11 terrorists attacked the World Trade Center and the Pentagon because these structures symbolized economic and military might, and they were recognizable throughout the world. Although commercial nuclear facilities such as power plants are not affiliated with the military, they, like the buildings struck on 9/11, signify economic and national power. Unlike the 9/11 buildings, however, nuclear facilities house sources of ionizing radiation. Some types of attackers would try to play on many people's fears of radiation exposure, especially if a successful attack released radioactive materials into the environment. A 9/11-scale attack on a nuclear plant could have a devastating psychological impact on the public and have a major financial impact on the nuclear industry. Other motivations for attack are to punish, intimidate, or blackmail the industry, government, or society.

Who would attack nuclear facilities?

Relatively few terrorist groups are motivated to carry out an attack on a nuclear facility. The list of suspects includes certain types of terrorist groups, people who are motivated by a political or environmental cause, saboteurs who may or may not work at the facility, and deranged people. While no terrorist group has attacked a nuclear power plant or related nuclear facility, some groups have expressed interest in or at least considered launching such an attack.

National separatist or national unity groups seek to liberate an occupied or oppressed territory or to unify that territory with another country. For example, the Irish Republican Army sought for many years to liberate Northern Ireland from British rule. The Basque separatists still want to free their lands from Spanish influence. For a recent example relevant to nuclear security, following the dissolution of the Soviet Union, the Chechen rebels aimed to separate Chechnya from Russian rule. This group demonstrated possession of radioactive materials and indicated that they could detonate a radioactive "dirty bomb" or radiological dispersal device, if they wanted to do so. In November 1995, for instance, Chechen rebels placed a container of radioactive cesium-137 in Ismailovsky Park in Moscow, and called a television crew to film the container. Because of the news media attention and the nondetonation of the radioactive material, many experts speculate that the rebels wanted to use this event for psychological purposes.

Another germane example relating to a nationalist group is Rodney Wilkinson, a white South African who was recruited by the African National Congress (ANC) because of his sympathies to the ANC's cause of eliminating the apartheid rule. In 1982, Wilkinson was working at the nearly commissioned Koeberg Nuclear Power Plant in South Africa. On December 18, he detonated four bombs that were designed to not cause too much damage on the reactor. In an interview with the *Guardian* of London, he claimed that the attack was timed on a Saturday when few people would risk being injured and was conducted prior to the operation of the plant to prevent the possible release of highly radioactive fission products. The intended effects were economic, symbolic, and psychological, in that the expensive plant was the first commercial nuclear reactor in the continent of Africa.

In general, terrorism experts assess that national separatist or national unity groups would not want to release radioactive materials that would contaminate their constituents' territory or to incite a massive retaliatory response against their constituents as a result of a radiological attack. First and foremost, these groups need the continuing support of their constituents and would not want to launch attacks that would unduly risk alienating their supporters.

Political-religious terrorist groups such as al Qaeda seek both political power and religious influence. Often, leaders of these groups hope for enough power to make their organizations as powerful as nation-states. Osama bin Laden, for example, has said he wants to establish a caliphate that would unify Muslim lands from Morocco to Indonesia. Nuclear weapons could help him achieve that goal. In fact, in 1998, bin Laden said it was "the religious duty" of al Qaeda to obtain weapons of mass destruction, including nuclear weapons. But would an attack on a nuclear power plant or other commercial nuclear facility help advance his political aims? Because such attacks would likely not cause massive numbers of fatalities—at least not in the near term, but in a worst case may increase cancer deaths over the long term—bin Laden may not rank this type of terrorism as a high priority. Nonetheless, some al Qaeda-affiliated groups may have interest in these kinds of attack because of the economic damage that might result.

Some radical environmentalists believe that nuclear power is morally reprehensible and is harmful to the earth. Because they want to protect the environment, they would not want to actually cause a release of radioactive materials, but they could try to disrupt the function of a nuclear plant. Perhaps the most severe action they could take is to cut electrical power lines from the plant. For example, in May 1989, members of the Evan Mechan Eco-Terrorist International Conspiracy (EMETIC) were charged with conspiring to damage power lines connected to the Central Arizona Project and the Palo Verde Nuclear Power Plant in Arizona, the Diablo Canyon Nuclear Power Plant in California, and the Rocky Flats Nuclear Facility in Colorado. EMETIC had splintered off from the radical environmental group Earth First!

Apocalyptic groups believe that one day an Armageddon or other tumultuous event will occur that will cleanse evil from the earth. The members of these groups believe themselves to be the chosen ones who will survive this event. While some apocalyptic groups are content to wait until the event happens, others seek to force it to occur. Aum Shinrikyo, an example of the latter, wanted to bring about the end of time. Shoko Asahara, the leader of this group, envisioned sparking a nuclear war involving the United States and Japan. The group was headquartered in Japan and had tried to obtain nuclear weapons. While this cult never acquired nuclear bombs, it made chemical weapons and experimented with biological weapons. Most infamously, cult members released sarin gas, a chemical that affects people's nervous systems, in five Tokyo subway cars on March 20, 1995. Although the delivery mechanism—puncturing sarin-filled polyethylene pouches with sharpened umbrella tips—was crude, twelve people died and more than 5,000 were injured. The cult intended to kill thousands with this and subsequent attacks. Fortunately, police pressure in 1995 and 1996 led to the group's undoing before Aum could launch more devastating chemical and biological attacks. On February 27, 2004, a Tokyo court sentenced Asahara to death for his crimes. According to the Study of Terrorism and the Responses to Terrorism, Aum Shinrikyo renamed itself Aleph in 2000, so as to shed its extremely negative image, but Aleph members were discovered gathering information on nuclear power plants. They had hacked into computer networks to acquire information about nuclear facilities in China, Japan, Russia, South Korea, Taiwan, and Ukraine.

Right-wing extremists, such as some white supremacists, are not considered likely to attack nuclear power plants. But some of them have found inspiration in *The Turner Diaries*. That book contains several scenes in which white supremacists use nuclear and radiological weapons. In particular, one fictional attack involves combining two aspects of radiological terrorism. That is, the terrorists use dirty bombs to attack a nuclear power plant to contaminate the grounds to make it unusable without its undergoing an expensive cleanup. The closing scenes of the book are riddled with many nuclear detonations in American cities. In the final scene, the protagonist is ordered to undertake a suicide mission by flying an airplane containing a nuclear bomb into the Pentagon. Fortunately, in the past decade, some terrorism experts have assessed a weakening of right-wing extremism in the United States. For instance, William Pierce, the author of *The Turner Diaries*, died in 2002, and several other leaders of the White Aryan movement have also died. Other terrorism experts, however, warn that a leadership void can create a power vacuum, which can result in further radicalization and extreme acts of violence to maintain the viability of the movement.

What are the potential modes of attack or sabotage, and what has been done to protect against them?

Because of the sensitive nature of this subject, only basic information that is already in the public domain and does not reveal specific vulnerabilities about any facility is presented here. The six means of attack or sabotage generally considered are airplane crashes, truck bombs, commando attacks by land, waterborne attacks, insider collusion, and cyber attacks.

Immediately after the 9/11 airplane attacks on the World Trade Center and the Pentagon, concerns were raised that terrorists might crash airplanes into nuclear power plants. The 9/11 attack planes were laden with flammable jet fuel. At that time, David Kyd, a spokesperson for the IAEA, admitted, "If you postulate the risk of a jumbo jet full of fuel, it is clear that their [nuclear power plants'] design was not conceived to withstand such an impact." According to the 9/11 Commission report, Mohammad Atta, one of the lead al Qaeda attackers, considered smashing an airplane into a target code-named "electrical engineering," which was believed to have referred to the Indian Point Nuclear Power Plant in upstate New York. Because it is located just fifty miles from New York City, this plant has received increased security attention. Atta was reportedly deterred from attacking this facility because of his worries about the restricted air space.

Attackers using trucks, vans, or other vehicles laden with explosives have had devastating effects on military bases, embassies, other government buildings, and commercial structures. In 1983, two truck-bomb attacks on American assets in Lebanon served as harbingers of greater damage to come. On April 18 of that year, 63 people were killed in the truck bombing of the American embassy in Beirut; and on October 25, another truck bomb killed 241 American military personnel at the Marine Corps barracks in Lebanon. The U.S. Nuclear Regulatory Commission (NRC) assessed whether to require additional protection against truck bombs at nuclear plants, but it decided to not issue this regulation. On February 26, 1993, a van laden with more than 1,000 pounds of explosives rocked the north tower of the World Trade Center. A federal investigation concluded that Ramzi Yousef, an affiliate of al Qaeda, had organized this attack. In response, the NRC required licensees to install truck-bomb barriers and to include this threat into the design-basis threat assessment. After the April 19, 1995, bombing of the Alfred P. Murrah federal building in Oklahoma City that killed 168 people, the NRC conducted a renewed examination of truck-bomb protections. These upgrades were completed by February 1996. The Oklahoma City attack was done by Timothy McVeigh and Terry Nichols, two American terrorists who sympathized with the anti-government American militia movement.

Even before the 9/11 attacks, the NRC and other nuclear regulatory agencies had included the possibility of a small commando-style attack in the design-basis threat assessments. Both governmental and nongovernmental security experts, however, have raised concerns that the design-basis threat assessment has assumed a commando force too small, as compared to the nineteen well-organized attackers working in four teams to carry out the 9/11 attacks. Further questions were raised about security procedures at the Sizewell B nuclear facility in Suffolk, England, after Greenpeace activists breached the security perimeter on October 14, 2002, and two security guards took 25 minutes to accost them.

Another debatable issue is how well armed the commandos would be. They would probably have access to automatic weapons such as machine guns and they may possess rocket-propelled grenades and other armor-piercing weapons. These attackers may, in addition, arrive in armored vehicles such as specially equipped sport utility vehicles. In the event of a large ground attack in the United States, security guards at a nuclear plant would try to fend off the attackers until backup forces such as the National Guard could show up. Immediately after 9/11, guards at U.S. nuclear power plants were working many hours of overtime because of new NRC requirements, and many guards were becoming disgruntled. In early 2003, the NRC issued guidelines to improve the working conditions and encourage hiring of more guards, as well as to require more realistic force-on-force simulations to make sure that the guards are well prepared.

Remember that nuclear power plants require external sources of cooling water. Often, lakes, oceans, or seas provide this heat sink. Terrorists could try to shut off access to the cooling water by blocking the water intakes. But this type of attack is highly unlikely to damage a plant, let alone cause a reactor-core meltdown. Nonetheless, extended blockage could have an economic impact. Reportedly, nuclear power plants have barriers around the water intakes to impede such an attack. The revised design-basis threat assessment that followed the 9/11 terrorist attacks is believed to require such barriers.

Probably the greatest concern is that one or more workers at a nuclear plant would help out some attackers. Soon after 9/11, NRC Chairman Richard Meserve called insiders the "most difficult threat to defend against." Insiders are most threatening because they have detailed knowledge of a plant's operations and they could get access to potentially vulnerable parts of the plant. They would likely be most effective when acting in concert with external attackers. For example, if an airplane crashes into and breaches the containment building, an insider could disable the emergency core-cooling systems to help create the conditions for release of radioactive materials into the environment. Design-basis threat assessments assume that at least one insider could work with external attackers. Consequently, owners of plants are required to conduct background checks of workers and to watch for any suspicious behavior of their workers. Moreover, owners can implement a two-person rule that would require more than one person be present to gain access to sensitive areas of a plant. Furthermore, after 9/11, the NRC tightened visitor controls to require escorted access throughout a plant in almost all circumstances.

Cyber attacks are an emerging vulnerability for nuclear plants. While cyber terrorism has been around for decades, nuclear plants had generally been relatively immune because the older plant designs used analog controls. But with the increasing use of digital control systems, especially in the newer plants, the potential for cyber attacks has increased. For example, in January 2003, the slammer computer worm penetrated a computer network at the Davis Besse Nuclear Power Plant in Ohio. The plant was vulnerable because information-technology personnel had not installed a Microsoft patch that had been available for about six months. Fortunately, the plant was not operating when infected. But even if it had been, a plant spokesperson said that analog devices would have served as a backup. Recognizing the increasing vulnerability to cyber threats, the NRC in 2002 began a research program to investigate improved defenses against such attacks.

What more can be done to strengthen the security of existing and future facilities?

In recent years, the nuclear industry has made a number of security improvements. Generally, nuclear power plants have in place defense-in-depth security systems so that attackers would have to defeat multiple layers of security to reach the vital areas of a plant. The training of guards has also improved. Plant owners have increased controls on access. Perhaps most important, authorities have revised the design-basis threat (DBT) assessment following the 9/11 attacks. Nonetheless, government inspectors and independent experts have raised concerns that this threat assessment needs further enhancement. Despite these concerns, in January 2009, the NRC commissioners narrowly voted down a staff recommendation to further revise the DBT. The vote was two to two, and the NRC voting rules stipulated that a tie vote disapproved a staff recommendation. (At that time, there were only four of the five commissioners appointed because the White House had yet to appoint the fifth commissioner who would have broken the tie vote.) Dale Klein, the NRC Chairman at that time, said that the vote could be reconsidered after completion of a U.S. government assessment of security measures. A few years earlier, the U.S. Government Accountability Office had recommended that the NRC "improve its process for making changes to the DBT and evaluate and implement measures to further strengthen its force-on-force inspection program." In particular, this program had appeared to give guards too much advance warning and information about a pending test. On the one hand, a surprise test may jeopardize plant operations; but, on the other hand, providing too many details of a planned force-on-force test would not adequately determine the capabilities of guard forces. The Project on Government Oversight, a watchdog organization, has interviewed hundreds of security officers and has concluded that "security culture is a real problem." Many of these officers have expressed concern that plant management does not take security seriously enough. Weak enforcement of security lapses can undermine security culture. For instance, the NRC fined Exelon only $65,000 for their guards' sleeping on duty. In comparison, the NRC spent almost $500,000 investigating this incident.

Safety and security teams can work more closely together to develop and implement protections against the vulnerabilities identified in the design-basis threat assessment. The DBT would benefit from frequent input from intelligence agencies to obtain the latest information about possible attackers. Regarding protection against airplane crashes, the NRC in February 2009 issued a rule to require applicants for new plants to assess the ability of these facilities to withstand the effects of an impact from a large commercial aircraft. The ruling made clear that the government has the responsibility to prevent aircraft hijacking. To foster other security improvements to future plants, security teams need to work side-by-side with plant design teams.

What military attacks have there been on nuclear reactors?

According to Bennett Ramberg, in *Nuclear Power Plants as Weapons for the Enemy: An Unrecognized Military Peril*, governments may be motivated to use their militaries to attack their enemies' reactors in order to damage their electrical power system, destroy a potent status symbol, impede the ability to make fissile material for nuclear weapons, or contaminate their enemies' territories with radioactive material. The Middle East stands out as a region that has experienced repeated military attacks on commercial nuclear power plants that were designed for electricity generation, as well as research reactors that were believed to have been built for production of plutonium to fuel nuclear bombs. For example, on September 30, 1980, Iran's air force bombed the Osiraq research reactor being built by France in Iraq. The power rating of this reactor was ideal for making at least one bomb's worth of plutonium every year of operation. But this Iranian attack did not destroy the Iraqi reactor. Israel decided to finish the task. On June 7, 1981, Israel launched an air attack that destroyed the reactor. Iraq did not rebuild this reactor, or any other reactor that could have supplied a bomb's worth of plutonium annually. Saddam Hussein instead learned that he needed to move his nuclear weapons program out of sight, and he thus pursued a covert uranium-enrichment program throughout the 1980s and into 1991, during the First Gulf War. Iraq's defeat in that war permitted international nuclear inspectors to have greater access to Iraq. They found that Saddam's nuclear scientists were getting close to making a uranium-based nuclear bomb.

Saddam also learned the lesson of stopping his enemy Iran from acquiring a plutonium-production capability. During the long and bloody Iran–Iraq War in the 1980s, he repeatedly targeted Iran's then-partially built Bushehr Nuclear Power Plant. On March 24, 1984, Iraqi warplanes first bombed this plant. Although the plant suffered light damage during this attack, the Iraqi military then struck six more times—twice in

1985, once in 1986, twice in 1987, and a final attack in 1988. These repeated bombings severely damaged one of the two reactors that were under construction. A more recent bombing of a suspected nuclear facility occurred in September 2007, when Israel destroyed a building in Syria. The U.S. government reported that Syria was building a research reactor of the same or similar design as the North Korean reactor at Yongbyon, which has produced weapons-grade plutonium. North Korean technicians were also reportedly seen at the Syrian site.

What can countries do to protect their nuclear facilities from military attacks?

Nations can deploy their air defense systems, harden their facilities, and reach agreements with potential attackers to place some or all of the facilities off-limits to attack. Yet, very few nations have taken these steps to protect their nuclear facilities. Iran has deployed air defenses and partially or fully buried some of its most valuable facilities, especially those that enrich uranium. In 1988, India and Pakistan signed an agreement to exchange annual lists of a selected number of their nuclear facilities. The agreement was ratified in 1990, and the exchange has taken place since 1991; the lists are not made public, and it is believed that not all facilities are listed. Nonetheless, this agreement serves as a confidence-building measure even though it does not guarantee that a military attack will not occur. Aside from these efforts, governments can help secure their nuclear facilities by maintaining adequate military forces and by forming security alliances with stronger states.

7

RADIOACTIVE WASTE
MANAGEMENT

**What are the types of radioactive waste,
and how are they generated?**

The three main types of radioactive waste are low level, inter-
mediate level, and high level. According to the U.S. Nuclear
Regulatory Commission, low-level waste "includes items
that have become contaminated with radioactive material or
have become radioactive through exposure to neutron radiation." This waste usually consists of materials emitting truly
low amounts of ionizing radiation—materials like protective
shoe covers, medical tubes, mops, rags, syringes, and contaminated laboratory-animal carcasses. However, despite
the name, low-level waste can also consist of materials that
are highly radioactive—in particular, parts from inside the
reactor pressure vessel of a power plant. Outside the United
States, many countries often designate such radioactive
materials from reactor vessels as intermediate-level waste,
which can also include filter ion-exchange resins, filter
sludge, precipitates, evaporator concentrates, incinerator
ash, and fuel cladding.

Usually, low-level waste with low amounts of radiation is
allowed to decay in storage locations at sites where the waste

was generated. Intermediate-level waste tends to require longer isolation from the environment to allow for sufficient radioactive decay. Short-decay materials have half-lives typically less than thirty years. Low-level waste is often accumulated into large enough shipments to be sent to a licensed low-level waste-disposal site. These shipments must use government-approved containers that meet safety and security requirements.

High-level waste is generated by the fission reactions inside reactors. While many of the components of this waste decay within a short period of time—from a matter of seconds to a few days—the remainder of the waste lasts decades, even up to tens of thousands of years. Because of the very potent radiation, handling high-level waste requires special procedures and sufficient amounts of shielding to protect workers' and the public's health.

What is the typical composition of spent nuclear fuel?

Every eighteen to twenty-four months, a fuel assembly is removed from a commercial light-water reactor. The removed material is called "spent," "irradiated," or "used" fuel; here, for consistency, the term of choice is "spent fuel." A spent fuel assembly will approximately consist of, by weight, 95.6 percent uranium; 0.9 percent plutonium; 0.1 percent minor actinides (americium, curium, and neptunium, for example); and 3.4 percent fission products, which are produced from fission of uranium and plutonium. Recall from chapter 1 that an atom of a fission product has, on average, half the mass of a uranium or plutonium atom. Actually, the distribution of fission product masses is such that about half of the atoms are somewhat less than half the mass of uranium or plutonium, and the other half of the fission products has masses slightly more than half (of) the mass of uranium or plutonium. Typically, 98 to 99 percent of the uranium is nonfissile uranium-238, with almost all the remainder being fissile uranium-235. The more recent practice of using higher burn-up fuels, which are designed to fission more of the uranium-235 than traditional fuels, has decreased the concentration of uranium-235 in spent fuel to less than 1 percent. Such practice increases efficient use of uranium resources but can make storage of spent fuel more challenging because of its larger amounts of highly radioactive fission products.

How long does the radioactivity in spent fuel last?

When a spent-fuel assembly is removed from a reactor, radioactive decay generates tens of kilowatts of heat, which means that the materials are rather hot and require cooling. Cooling is provided by placing the spent fuel in large and deep pools of water. The water is deep enough to offer a shielding layer of several meters above the spent fuel. This shielding is sufficient to allow workers to walk around the top of the pool without experiencing radiation exposure from the spent fuel. Spent fuel has to reside in the pool for at least a few years in order for the radioactivity to decay to low enough levels to permit removal from the pool. In particular, after five years, the heat from the radioactive decay has decreased by a factor of about 100. At this time, the spent fuel could be placed in dry storage casks. But because of the relatively high expense of these casks, often power plant owners decide to keep spent fuel in the pools for a longer period as long as there is enough room. As indicated by the factor of 100 radioactivity reduction after five years, most of the radioactive substances in spent fuel have short half-lives of less than one year. For the next few hundred years, the substances of greatest concern are cesium-137 and strontium-90, which have about thirty-year half-lives. After a few tens of thousands of years after the creation of the spent fuel, the radioactive materials have decayed to the level of the original ore containing the uranium for the fresh fuel. But the concentration of these materials is much greater in spent fuel than in ore. Moreover, the spent fuel still contains plutonium-239 and a few other radioactive substances, even after tens of thousands of years.

How hazardous is radioactive waste?

The hazardousness of radioactive waste depends on the half-life of the radioactive substance, the type of ionizing radiation emitted, the energy content of the ionizing radiation emitted by that substance, and the possible pathways for that substance to enter the environment, especially the food chain or water supplies. A short half-life means that the substance will decay quickly; after seven half-lives, less than 1 percent of the original substance will remain. The decay products are often radioactive, but the decay chain eventually ends at a stable substance. As discussed in chapter 1, the three main types of ionizing radiation are alpha, beta, and gamma, which range from least to most penetrating. Alpha emitters pose internal health hazards only if ingested or inhaled. Beta emitters may present an external health hazard, especially for unprotected eyes, if the beta radiation is highly energetic. Gamma emitters always pose both internal and external hazards. In addition to properties of nuclear decay such as radioactive half-life and type of radiation, there are chemical properties that affect the behavior of the radioactive substance in the environment or the human body. For instance, substances that are water soluble may pose a significant hazard if large enough quantities migrate into water supplies and may enter the food chain. Strontium-90, a beta emitter and a fission product, for example, tends to become incorporated in bones because it behaves chemically like calcium, an elemental building block of bones.

How much spent nuclear fuel has been produced?

About 270,000 metric tons of spent fuel are in storage worldwide. Most of the spent fuel is stored on-site at the power plants. Of this, approximately 90 percent is contained in pools of water. Every year, about 12,000 metric tons of spent fuel is discharged from about 440 commercial reactors. Roughly, one-fourth, or about 3,000 metric tons, are sent to reprocessing facilities. Because the United States has the largest number of reactors, it generates the largest proportion of the world's spent fuel. Annually, about 2,000 metric tons of spent fuel is discharged from 104 U.S. reactors. Cumulatively, the United States has about 60,000 metric tons of this material. The vast majority of it is stored in pools, and a small portion has been placed in dry storage casks. None of it has been sent to permanent storage.

What are the storage options for dealing with radioactive waste?

The general principle is to keep exposure to people as low as reasonably achievable. Minimizing the exposure time, maximizing the distance from the radiation source, and ensuring adequate shielding are the three fundamentals for radiation protection. Isolating radioactive waste from the environment can be done by storing it in well-sealed containers that can withstand attack and conceivable stressful events, including intense fires and impacts such as accidental crashes and intentional attacks. As mentioned above, pools for spent fuel provide immediate protection; dry storage casks offer safe and secure storage for several decades.

Longer term, the preferred option for storing radioactive waste has been to use underground disposal in geological repositories, which are mined deep into mountains or deep below the earth's surface. Multiple barriers have been designed to prevent radioactive waste from leaking from waste storage. For high-level waste separated out during reprocessing, the radioactive materials are mixed and immobilized in glass, producing what is called "vitrified waste." The waste is placed in canisters that are then stored in shielded high-level waste-storage facilities. Nonreprocessed spent fuel is intended to be sealed inside corrosion-resistant containers made of stainless steel or copper. These containers can then be placed in deep underground repositories. The geological formations of these locations would also provide natural barriers to the leakage of radioactive material. In particular, non-porous rocks would help prevent the diffusion from the repository of liquid radioactive material that might leak from the containers. In addition, drip shields, a man-made barrier, can be included in the repository to prevent or at least reduce the amount of moisture in contact with the containers of spent fuel. Moreover, bentonite clay can be used as an impermeable backfill for repositories subject to moisture.

What is the volume of radioactive waste, and how does this compare to other industrial toxic waste?

Radioactive waste takes up very little volume compared to the total volume of toxic waste produced by industry. According to the World Nuclear Association, less than 1 percent of the total volume of toxic waste is made of nuclear waste. About 90 percent of the radioactive waste is low level, but this is only about 1 percent of the total radioactivity. Conversely, high-level waste takes up a small fraction of the total volume of toxic waste, but is about 95 percent of all produced radioactivity. Annually, nuclear power globally generates approximately 200,000 cubic meters of low- and intermediate-level wastes and about 10,000 cubic meters of high-level waste, including spent fuel. For a typical 1,000-MWe reactor, every year it produces 200 to 300 cubic meters of low- and intermediate-level wastes and 20 cubic meters of spent fuel. This spent fuel weighs about 27 metric tons. Once it is placed in a storage container, the volume is about 75 cubic meters. If the spent fuel is reprocessed, the vitrified waste of highly radioactive fission products takes up about 3 cubic meters, which fills up 28 cubic meters when placed in a storage canister. In comparison, a 1,000-MWe coal-fired power plant would release about 400,000 metric tons of ash annually.

How vulnerable are spent fuel pools?

A 2004 U.S. National Academy of Sciences (NAS) study concluded that "successful terrorist attacks on spent fuel pools, though difficult, are possible." According to the study, "If an attack leads to a propagating zirconium cladding fire, it could result in the release of large amounts of radioactive materials." But the cladding would burn only if it was exposed to air, and thus the water would have to be drained from the pool. Doing so would be very challenging for a terrorist group to accomplish. This study was done because nongovernmental experts expressed concern in 2003 about the vulnerability of these pools. They had recommended removal of the older spent fuel from the pools to be placed in dry storage casks, which were deemed more secure.

But buying and loading each storage cask cost more than $1 million, and three to four casks are needed to hold the spent fuel from each reactor discharge. An additional cost would involve construction of a facility to house the dry storage casks. So, a plant owner would have to pay potentially tens of millions of dollars to transfer to dry storage casks the older spent fuel that had been stored over many years in the pools. Moreover, the transfer to dry storage casks raises the risk of workers' exposure to radiation. The NAS study did not endorse this recommendation because less expensive methods of storage are available. In particular, the study endorsed rearranging the spent fuel in the pool, thereby surrounding the hotter, newly discharged assemblies with the cooler, older assemblies. Doing so would likely prevent the propagation of a fire. In addition, a water spray system may mitigate the potential vulnerabilities. But this system may be justified only after a plant has undergone a thorough cost versus benefit analysis.

Does reprocessing reduce the amount of nuclear waste?

This question was considered in chapter 2, on energy security, and it was pointed out that reprocessing does significantly reduce the volume of high-level nuclear waste with the removal of uranium and plutonium and the nonradioactive fuel cladding, although reprocessing generates additional low-level waste. The separated radioactive fission products take up much less space than the other materials. But these fission products are much more radioactive than uranium or even plutonium. So, reprocessing does not significantly reduce the radiation hazards, though it does concentrate the high-level waste into smaller volumes. This high-level waste requires storage in safe and secure locations.

How do nuclear plants and coal plants compare in terms of radioactivity emitted?

One of the arguments in favor of nuclear power is that an operating nuclear power plant does not emit any radioactivity as long as it is safely run. By comparison, coal-fired power plants emit radioactive materials in the fly ash. These materials are made up of uranium and other naturally occurring radioactive substances that are present in the coal. According to the Oak Ridge National Laboratory, fly ash releases 100 times more radioactive material than does a nuclear plant for the same amount of electrical energy produced. Unless trapped, this ash is emitted into the environment. Fly ash can also seep into water supplies and the food chain. The Oak Ridge study assessed that a person living near a coal plant receives a maximum exposure of 1.9 millirems of radiation from the ash. But this pales in comparison to the average exposure of 360 millirems from background radiation. Thus, the radiation risk from a coal plant is very small. This is not to say that coal plants do not have major hazards, however. They emit copious quantities of carbon dioxide—a greenhouse gas—and worldwide, thousands of coal miners perish or suffer substantial harm to health. Moreover, certain mining practices, such as mountain-top removal, a form of strip mining, have potentially devastating environmental effects. Further, coal plants that do not capture the sulfur dioxide and nitrous oxides exacerbate acid rain, which has had a devastating effect on forests, especially in the eastern United States.

A cap-and-trade system for limiting emissions of these gases was established in the 1990s, and has been successful in substantially reducing acid rain. These emissions—particularly sulfur dioxide—can exacerbate respiratory conditions such as asthma. Nuclear plants do not emit these gases.

What country is closest to opening a permanent nuclear waste repository?

No country has opened a permanent nuclear-waste repository. But a few countries, such as Finland and Sweden, have made some significant progress toward this goal. Sweden, in particular, appears to be closest to opening a repository around the year 2020. The Swedish government decided early on to make the process of selecting a repository as transparent as possible and to involve many diverse interest groups. For instance, Greenpeace and other anti-nuclear groups were invited and took part in discussions. Local citizens provided their input. Perhaps the smartest move on the part of the Swedish authorities was to consider three different sites and then narrow down to two finalists. These sites were all thoroughly assessed, giving authorities more than one option and providing confidence that a particular locale was not a foregone conclusion. All site communities were volunteers and did not have to accept the facility if they decided against it. Many in these communities began to see that hosting a repository, if managed safely and securely, could offer a significant number of jobs.

How hazardous is the transportation of radioactive waste?

The nuclear industry has transported radioactive waste for more than fifty years, with no major mishaps. However, the quantities that have been shipped have been small compared to the huge amounts that are planned to be sent to permanent repositories. In particular, only approximately 3,000 metric tons of the total 60,000 metric tons of the U.S. inventory have been moved, though not to a permanent repository. This inventory is spread out over about seventy sites. Other countries are in a similar situation, but the scale of the U.S. nuclear-waste issue is greater than other countries, as the United States has the largest nuclear power program. An authoritative study by the U.S. National Research Council's Nuclear and Radiation Studies Board and the Transportation Research Board concluded that "transport by highway (for small-quantity shipments) and by rail (for large-quantity shipments) is, from a technical standpoint, a low radiological risk activity with manageable safety, health, and environmental consequences when conducted in strict adherence to existing regulations." But the study underscored the "social and institutional challenges" to moving tens of thousands of tons to a permanent repository. The experts recommended detailed examination of proposed transportation routes for any hazards that could result in extreme accidents. Such accidents would involve very long and intense fires that engulf casks of spent fuel. By comparison, "normal" highway crashes and less than extreme fires would not breach the storage containers or release radiation, based on numerous tests. People along these routes may fear radiation exposure; thus, the study emphasized addressing social risks such as potential "reductions in property values . . . reductions in tourism, increased anxiety, and stigmatization of people and places." Another concern is malevolent acts, such as sabotage or terrorist attacks, necessitating protection by security forces.

Why was Yucca Mountain chosen as the permanent repository in the United States, and what will be its fate?

In 1986, the U.S. Congress passed an amendment to the radioactive-waste storage act that named Yucca Mountain in Nevada as the only repository site to be considered. Nevada has never had a commercial nuclear power plant. While the United States had been investigating more than one waste repository, Nevada had a weak congressional delegation in the late 1980s and was not able to prevent Yucca Mountain from becoming the sole site. The congressionally mandated storage limit at that site was 70,000 metric tons until a second repository would be in operation; if that occurred, then the Yucca facility would be allowed to be expanded by law, though it might face political resistance.

By 2020, the U.S. inventory of cumulative spent fuel will have exceeded that congressional limit. Some technical studies, however, indicate that the site could hold much more spent fuel. The Department of Energy has estimated that the site's capacity could exceed 120,000 metric tons. Moreover, according to an Electric Power Research Institute study, Yucca Mountain could contain at least four times the legislative limit—and possibly nine times that limit—allowing that site to store "all the waste from the existing U.S. nuclear power plants, but also waste produced from a significantly expanded U.S. nuclear power plant fleet for at least several decades."

But political opposition has cast doubt on this site's viability. By 2008, with the Democratic Party in the majority in both houses of Congress, and with Nevada Senator Harry Reid as majority leader, there was expressed desire to close down the Yucca Mountain project. In January 2010, President Barack Obama and Secretary of Energy Steven Chu, while having stated that Yucca Mountain is not workable, appointed a Blue Ribbon Commission of experts to assess other disposal options. While Yucca Mountain and other disposal options have been well studied from a technical standpoint, this commission may identify both politically and technically viable action plans. This commission is slated to publish its report by summer 2011.

Will delays in opening a permanent repository for radioactive waste derail continued or expanded use of nuclear power in the United States?

Continued delay in solving this waste-disposal problem will likely erode confidence in nuclear power. Industry and government have a responsibility to protect the public. The U.S. Nuclear Regulatory Commission, for instance, is required to issue a waste-confidence rule that shows a clear path forward for safe and secure waste disposal. The U.S. federal government, in particular, is obligated by law to establish a permanent repository. Because of the likelihood that Yucca Mountain will not be approved for use, and that the United States will not have a repository for at least a couple of decades, the utilities have begun to sue the government for breach of contract and have requested return of money paid into a several-billion-dollar fund for waste disposal. Electricity consumers have been paying into the fund via a 0.1 cent per kilowatt-hour fee on nuclear-generated electricity.

Despite the political impediments to using Yucca Mountain, the waste-storage problem in the United States can be managed. A possible pathway is to pursue a dual-track approach: develop a consensus to open up more than one permanent repository (which may or may not include Yucca

Mountain), and store as much spent fuel as possible in safe and secure dry casks at existing reactor sites. As in the original conception of permanent disposal, there would likely be two or more regional repositories for optimizing political fairness and minimizing the transportation risks. The federal government would likely have to provide money from the waste-disposal fund to cover the interim storage in the dry casks. This combination of interim storage and commitment to creating permanent repositories would probably provide the assurances needed by the public and the investment community for continued use of nuclear power. Long-term storage sites may want to preserve the option of retrieving spent fuel in case reprocessing can meet the conditions of safe, secure, and sustainable use of nuclear power. Many countries are already factoring this feature into their repository plans.

8

SUSTAINABLE ENERGY

What is meant by a "sustainable energy system"?

"Sustainability" can be described as "development that meets
the needs of the present without compromising the ability of
future generations to meet their own needs," according to the
Brundtland Commission, headed by former Norwegian
Prime Minister Gro Harlem Brundtland. But the amount of
energy consumed depends not just on people's survival needs
but also on the demands for continued economic growth, as
well as lifestyle choices that favor acquiring more and more
consumer goods. A world with the consumption pattern of
Europe, for example, would require roughly half the energy
per person as a country with the consumption pattern of the
United States. Consequently, in the debate over the future of
energy policy, in general, and nuclear energy policy, in particular, people need to differentiate between what is abso-
lutely needed to ensure a decent standard of living and what
is considered going beyond that standard.

These are the two major role models for the developing
world. Of the 6.8 billion people worldwide, 1.6 billion—
residing in the developing world—have very limited or

no access to electricity. Thus, the developing world's nations certainly require more energy to meet their economic needs. The energy choices they make now and in the foreseeable future will profoundly influence international security, the environment, and human health for decades, if not, centuries to come. The sustainable energy pathway seeks to create stronger security systems among nations, minimize detrimental impacts on the earth's ecosystem, and improve living standards.

What is a "renewable energy source"?

"Renewable energy" means that the source of energy can be replenished or not depleted such that it can be used by future human generations. For example, biofuels such as ethanol or biodiesel are sources of transportation fuel that can be replenished by growing more plants. Such sources are sustainable in the long term as long as the energy used to replenish them comes from renewable sources. That is, if fossil fuels are needed to make fertilizer for growing the biofuel plants, to transport the harvested plants to biofuel production facilities, and to move the ethanol and biodiesel to filling stations, then the overall system is not sustainable. But fossil-fuel usage may serve as a bridge to a fully sustainable system in which fossil fuels are eventually phased out.

Wind and solar energies are often described as renewable. But these sources will not last forever. Solar energy is available only while the sun is active. Wind energy results from solar heating of the atmosphere. Solar physicists estimate that the sun has probably another 5 billion years of hydrogen for fusion fuel. When the sun exhausts this fuel supply, it will switch to helium fusion and will begin to swell such that the outer part of the sun will eventually engulf the earth. Then, obviously, the earth will be uninhabitable, and the sun will not provide more useful solar energy. Five billion years is a very long time, and most likely humanity will be extinct by then or will have moved on to other locations in the universe to make use of other suns' energy sources. Thus, solar energy can be considered renewable for all practical purposes.

Is nuclear energy a renewable energy source?

Based on the definition above, nuclear energy that uses fissionable and fusion material existing on earth is not renewable. The earth-bound amounts of uranium and thorium, which can power fission reactions, are large but finite; similarly, deuterium, the heavy hydrogen needed for fusion reactions, is abundant in the world's water but is not inexhaustible. While the uranium, thorium, and heavy hydrogen cannot be replenished, in the way plants are for biofuels, it is possible to greatly extend fission fuels. As discussed earlier, breeder reactors, for example, can make plutonium from uranium-238. And thorium is a fertile material that can be used to produce uranium-233, a fissile isotope. In principle, because fertile uranium-238 is more than ninety-nine times more plentiful than fissile uranium-235, the supplies of fission fuel can greatly increase such that humanity could have thousands of years of nuclear energy. But the trade-off for doing so is to promote a plutonium fuel economy.

Can nuclear energy contribute to developing sustainable energy systems?

While a 2007 European Commission study on nuclear power and sustainability found in favor of nuclear power's being sustainable, this debate is far from settled. In particular, this study recognized that nuclear power could fulfill long-term energy needs only as long as reprocessing of spent fuel and fast neutron reactors were employed. These reactors would both consume long-lived radioactive waste and produce more plutonium from nonfissile uranium-238. The new plutonium would greatly extend nuclear fuel supplies because more than 99 percent of natural uranium is uranium-238. But this activity would result in widespread commerce in plutonium, potentially raising proliferation risks to unacceptable levels. Proponents for reprocessing argue that the reprocessing facilities and breeder reactors should be limited to countries that pose little or no proliferation risk. Presently, these efforts are largely confined to the existing nuclear-armed states and non-nuclear-armed state Japan, which is a U.S. ally. If use of reprocessing increases, more nonnuclear-armed countries would likely express interest in it. So, greater efforts in addressing security concerns are needed to reduce the proliferation risk.

Can renewable energies compete with nuclear and other base-load electrical power sources?

Presently, nuclear power offers a comparative advantage in providing reliable, base-load electrical power. As defined earlier, "base-load power" is the constant demand for electricity throughout the day and night. Power demands above base-load are called "peak power." Nuclear power plants are optimal for providing base-load because they are designed to be run at full power for several months. Coal-fired plants are also effective base-load sources. While plants using natural gas have also supplied base-load power, they are ideal for peak power because of their ability to be turned on quickly to meet the extra demand.

Nonhydro renewable energies such as solar and wind are viewed as intermittent sources. When the sun doesn't shine or is blocked by clouds, solar power plants will not generate power, and when the wind does not blow or blows at less than optimal speeds, wind turbines are not working at their best. This intermittency problem could change if storage systems for energy could allow for consistent power production from solar and wind power systems. Also, as indicated by a 2009 study by the research team led by Willett Kempton, linking enough wind farms may allow for base-load power generation from this energy source. The study examined five years' worth of wind data from eleven meteorological stations along a 2,500-kilometer stretch of the U.S. East Coast, simulating the link-up of large wind farms located hypothetically at each station. The study found that "the entire set of generators rarely reaches either low or full power, and power changes slowly." That is, the entire generating system would function as a very large base-load source. Wind power could, if fully harnessed at available locations, meet several times the world's present electricity demands. Moreover, smart electrical grids may offer the ability to more effectively use intermittent, renewable sources to provide reliable electricity.

As renewable and nuclear technologies continue to develop, the world may be generating much of its electricity from these sources by the century's end. Thinking even longer term, nuclear power may eventually serve as a bridging technology to a fully renewable energy future. Alternatively, nuclear power may experience widespread deployment in many countries over many centuries, as long as humanity remains vigilant in ensuring safe and secure use of peaceful nuclear energy.

SUGGESTIONS FOR FURTHER READING

Nuclear Energy: General Interest

Bernstein, Jeremy. *Plutonium: A History of the World's Most Dangerous Element.* Washington, D.C.: Joseph Henry, 2007.

Bodansky, David. *Nuclear Energy: Principles, Practices, and Prospects,* 2nd. ed. Woodbury, N.Y.: Springer, 2008.

Caldicott, Helen. *Nuclear Energy Is Not the Answer.* New York: New Press, 2006.

Cravens, Gwyneth. *Power to Save the World: The Truth about Nuclear Energy.* New York: Vintage, 2008.

Domenici, Pete V., with Blythe J. Lyons and Julian J. Steyn. *A Brighter Tomorrow: Fulfilling the Promise of Nuclear Energy.* Lanham, Md.: Rowman & Littlefield, 2004.

Garwin, Richard L., and Georges Charpak. *Megatons and Megawatts: A Turning Point in the Nuclear Age?* New York: Knopf, 2001.

Herbst, Alan M., and George W. Hopley. *Nuclear Energy Now.* Hoboken, N.J.: Wiley, 2007.

Josephson, Paul R. *Red Atom: Russia's Nuclear Power Program from Stalin to Today.* New York: W.H. Freeman, 2000.

Makhijani, Arjun. *Carbon-Free and Nuclear-Free: A Road Map for U.S. Energy Policy.* Takoma Park, Md.: IEER Press, 2007.

Marples, David R., and Marilyn J. Young, eds. *Nuclear Energy and Security in the Former Soviet Union.* Boulder, Colo.: Westview Press, 1997.

Morone, Joseph G., and Edward J. Woodhouse. *The Demise of Nuclear Energy? Lessons for Democratic Control of Technology.* New Haven and London: Yale University Press, 1989.

Morris, Robert C. *The Environmental Case for Nuclear Power: Economic, Medical, and Political Considerations.* St. Paul, Minn.: Paragon House, 2000.

Reynolds, Albert B. *Bluebells and Nuclear Energy*. Madison, Wisc.: Cogito Books, 1996.

Tucker, William. *Terrestrial Energy: How Nuclear Energy Will Lead the Green Revolution and End America's Energy Odyssey*. Savage, Md.: Bartleby Press, 2008.

Wilson, P. D., ed. *The Nuclear Fuel Cycle: From Ore to Waste*. New York: Oxford University Press, 1996.

Wolfson, Richard. *Nuclear Choices: A Citizen's Guide to Nuclear Technology*. Cambridge, Mass.: MIT Press, 1993.

Nuclear Safety and Waste Disposal

Macfarlane, Allison M., and Rodney C. Ewing, eds. *Uncertainty Underground: Yucca Mountain and the Nation's High-Level Nuclear Waste*. Cambridge, Mass.: MIT Press, 2006.

National Research Council. *Risk and Decisions about Disposition of Transuranic and High Level Radioactive Waste*. Washington, D.C.: National Academies Press, 2005.

Perin, Constance. *Shouldering Risks: The Culture of Control in the Nuclear Power Industry*. Princeton, N.J.: Princeton University Press, 2005.

Perrow, Charles. *Normal Accidents: Living with High-Risk Technologies*. Princeton, N.J.: Princeton University Press, 1999.

Rees, Joseph V. *Hostages of Each Other: The Transformation of Nuclear Safety Since Three Mile Island*. Chicago: University of Chicago Press, 1994.

Vandenbosch, Robert, and Susanne E. Vandenbosch. *Nuclear Waste Stalemate: Political and Scientific Controversies*. Salt Lake City, Utah: University of Utah Press, 2007.

Walker, J. Samuel. *Three Mile Island: A Nuclear Crisis in Historical Perspective*. Berkeley: University of California Press, 2004.

Walker, J. Samuel. *The Road to Yucca Mountain: The Development of Radioactive Waste Policy in the United States*. Berkeley: University of California Press, 2009.

Nuclear Proliferation

Albright, David. *Peddling Peril: How the Secret Nuclear Trade Arms America's Enemies*. New York: Free Press, 2010.

Campbell, Kurt M., Robert J. Einhorn, and Mitchell B. Reiss, eds. *The Nuclear Tipping Point: Why States Reconsider Their Nuclear Choices*. Washington, D.C.: Brookings Institution Press, 2004.

Cirincione, Joseph. *Bomb Scare: The History and Future of Nuclear Weapons*. New York: Columbia University Press, 2007.

Doyle, James E., ed. *Nuclear Safeguards, Security, and Nonproliferation: Achieving Security with Technology and Policy*. Amsterdam: Butterworth-Heineman, 2008.

Leventhal, Paul, Sharon Tanzer, and Steven Dolley, eds. *Nuclear Power and the Spread of Nuclear Weapons: Can We Have One Without the Other?* Dulles, Va.: Brassey's, 2002.

Mozley, Robert F. *The Politics and Technology of Nuclear Proliferation.* Seattle and London: University of Washington Press, 1998.

Mueller, John. *Atomic Obsession: Nuclear Alarmism from Hiroshima to Al-Qaeda.* New York: Oxford University Press, 2010.

Quinlan, Michael. *Thinking About Nuclear Weapons: Principles, Problems, and Prospects.* New York: Oxford University Press, 2009.

Reiss, Mitchell. *Bridled Ambition: Why Countries Constrain Their Nuclear Capabilities.* Washington, D.C.: Woodrow Wilson Center Press, 1995.

Rhodes, Richard. *Arsenals of Folly: The Making of the Nuclear Arms Race.* New York: Knopf, 2007.

Sokolski, Henry D. *Best of Intentions: America's Campaign Against Strategic Weapons Proliferation.* Westport, Conn. and London: Praeger, 2001.

Sokolski, Henry, ed. *Gauging U.S.-Indian Strategic Cooperation.* Carlisle, Pa.: Strategic Studies Institute, 2007.

Younger, Stephen M. *The Bomb: A New History.* New York: HarperCollins, 2009.

Zarate, Robert, and Henry Sokolski, eds. *Nuclear Heuristics: Selected Writings of Albert and Roberta Wohlstetter.* Carlisle, Pa.: Strategic Studies Institute, 2009.

Nuclear Terrorism and Military Attacks on Nuclear Facilities

Allison, Graham. *Nuclear Terrorism: The Ultimate Preventable Catastrophe.* New York: Times Books, 2004.

Cameron, Gavin. *Nuclear Terrorism: A Threat Assessment for the 21st Century.* Basingstoke, UK: Palgrave MacMillan, 1999.

Ferguson, Charles D., and William C. Potter, with Leonard S. Spector, Amy Sands, and Fred L. Wehling. *The Four Faces of Nuclear Terrorism.* New York: Routledge, 2005.

Jenkins, Brian Michael. *Will Terrorists Go Nuclear?* Amherst, N.Y.: Prometheus, 2008.

Levi, Michael A. *On Nuclear Terrorism.* Cambridge, Mass.: Harvard University Press, 2007.

Ramberg, Bennett. *Nuclear Power Plants as Weapons for the Enemy: An Unrecognized Military Peril.* Berkeley: University of California Press, 1984.

INDEX

9/11 Commission Report, 181
accidents, 68, 137–38, 140–43,
   157–58, 162, 175, 199
   Chernobyl, 28, 43, 138, 140,
     142–43, 149–53, 155,
     157–58, 167
   criticality, 140
   loss-of-coolant accident (LOCA),
     140–41, 146, 159
   Three Mile Island, 61, 138, 140,
     143, 146–48, 156–58, 167
acid rain, 73, 99, 197–98
Additional Protocol. See Model
   Additional Protocol to
   Comprehensive Safeguards
Afghanistan, 133
African National Congress (ANC),
   177
air travel, 23, 68, 157–58, 180–81,
   184, 186
Alaska, 89
Aleph, 180
Alfred P. Murrah Federal
   Building, 182
Alfvén, Hannes, 32
Algeria, 64
Al Qaeda, 133, 178, 181–82
alternative fuel cycle, 38
American Centrifuge Plant, 94
American Physical Society, 131
annealing, of reactor vessel, 139
Antarctica, 89

Araki, Hiroshi, 169
Areva, 66, 73, 76, 83
arms race, 103, 105, 112
Arrhenius, Svante, 88
Asahara, Shoko, 133, 179, 180
Atomic Energy Commission, 167
atomic vapor laser isotope
   separation (AVLIS). See laser
   enrichment
Atoms for Peace Program, 32, 118
attacks, 15–16, 67, 68, 108, 115,
   119, 129, 132–33, 136, 173–88,
   193–95, 199
   cyber, 181, 184, 185
   military, 115, 186, 187, 188
   nuclear facilities, 173, 176, 180,
     188
   Oklahoma City, 182
   radiological, 16, 177–78, 180
   September 11, 2001, 119
   terrorist, 32, 83, 103, 132–36,
     176–83, 195, 199
Aum Shinrikyo, 133, 179, 180
Australia, 37, 64, 79, 133
Austria, 20, 58, 152

Basque separatist, 132, 177. See also
   ETA
batteries, 17, 26, 52
Becquerel, Henri, 12, 25, 26
Belarus, 114, 152
Belgium, 56, 65

Beria, Lavrenti, 104
Berzelius, Jons Jakob, 38
Beyea, Jan, 168
Big Bang, 24
binding energy, 9, 10
Bin Laden, Osama, 133, 178
biofuels. *See* renewable energy source
biomass fuel, 93, 97
black market, nuclear, 36–37
Blue Ribbon Commission, 201
boiling water reactor, Mark I, 167–68
bombs
    dirty bomb, radiological, 16, 177, 180
    gun-type device, 104, 113, 134–35
    hydrogen, 105
    implosion-type device, 104
    improvised nuclear device (IND), 134
    nuclear, 11, 16, 22–23, 26, 28, 32–33, 38–39, 104–6, 110, 113, 121, 173, 179–80, 187
boric acid, corrosion of a reactor, 143
Brazil, 54–56, 64–65, 97, 117, 119, 131
Brundtland, Gro Harlem, 203
Brundtland Commission, 203
Bulgaria, 58, 59, 65, 143, 155
Bulletin of the Atomic Scientists, 110
bullwhip effect, of supply and demand, 74–75
Bush, George W., 84, 99, 101, 122, 127
Bushehr Nuclear Power Plant, based in Iran, 187

Canada, 44, 64, 65, 79, 89, 125–26, 153
Cancer, 13–14, 26, 27, 151, 178
    lung, 27
    thyroid, 151
cap-and-trade system, 72–73, 198
carbon sequestration, 92, 97
carbon tax, 72–73
Carter, Jimmy, 83
cascade, linking of centrifuges for uranium enrichment, 36
Chadwick, James, 19
chain reaction, 22–23, 28, 30, 49, 104, 140, 149
Cheney, Richard, 84

Chernobyl Nuclear Power Plant, 28, 43, 138, 140, 142–43, 149–53, 155, 157–58, 167
China, 18, 57, 64–65, 70–71, 76, 79–80, 100, 105–7, 109, 112, 115, 162, 170, 180
China National Nuclear Corporation, 162
China Syndrome, The, 146
Chu, Steven, 200
Churchill, Winston, 54
Clean Air Act, 73
Clean and Safe Energy Coalition, 101
Clean Development Mechanism (CDM), 100
clean energy, 98–99
climate change, 86, 88–91, 98–102
Clinton, Hillary, 109
Clinton administration, 99
coal, 5, 15–16, 54, 56, 60, 61–62, 68–70, 91–92, 94–96, 98, 162, 195, 197, 206. *See also* fossil fuels
Cochran, Thomas, 123
compounds, chemical
    chlorofluorocarbon, 87
    hydrocarbon, 5, 16
    methane, 5, 16–17, 86–87, 89, 92, 99
    nitrous oxide, 87, 99, 197
    potassium iodide, 151
    sulfur dioxide, 99, 197–98
    uranium hexafluoride, 34
    uranium tetrachloride, 33
confidence building measure, a means to prevent attacks on nuclear facilities, 188
Congress, U.S., 200
conservation, energy, 42
Constellation Energy, 72
consumption pattern, 203
contamination, radioactive, 27, 151–52, 162
coolant, for nuclear reactor, 29–30, 41–44, 46–50, 139, 149–50, 154, 160
Curie, Marie and Pierre, 26

Davis Besse Nuclear Power Plant, 144, 184
decay heat, 30, 140, 160

defense-in-depth, 138, 144, 160, 175,
    185
De Klerk, F. W., 113
Democritus, 7
Denmark, 100
design basis threat (DBT), 174, 182–86
deterministic effect, type of
    radiation effect, 14
deterrence, 115, 124
Diablo Canyon, 179
Diaz, Nils, 159
diffusion
    method of enrichment, 33–36, 94
    spreading of nuclear waste, 194
disposal
    nuclear waste, 15, 51, 81–82, 85,
        195–96, 198, 199
    toxic waste, 194–95
dual track approach, for radioactive
    waste disposal, 201
dual use, of nuclear technology, 32,
    37, 117, 120, 130

Earth First!, 179
earthquake, hazard to nuclear
    facilities, 163
Economic Simplified Boiling
    Water Reactor (ESBWR), 160
effusion, principle involved in
    uranium enrichment using
    diffusion method, 34
Egypt, 64, 119, 122
Einstein, Albert, 9, 21–22
Eisenhower, Dwight, 118
ElBaradei, Mohamed, 123
Electricité de France (EDF), 72
electric-powered vehicles, 56, 63
Electric Power Research Institute,
    200
electrolysis, 19
electromagnetic isotope separation
    (EMIS), 33, 34
electrons, 8, 12–13, 17
electron volts. See units of energy
elements. See Periodic Table of the
    Elements
embrittlement, of nuclear reactor
    vessel, 139, 154
emergency diesel generators, 144,
    149, 163, 164, 168, 169

emergency responders, 151
Eminent Persons Panel, advisory
    panel to the International Atomic
    Energy Agency, 124
energy
    chemical, 3, 5, 16
    frozen, 3, 21
    geothermal, 7, 68
    kinetic, 3, 10, 42, 43
    solar, 5–7, 56, 60, 62, 73, 91,
        204–5, 207
    wind, 6–7, 56, 60, 62, 73, 91,
        204, 207
enrichment, of uranium
    centrifuge method, 33, 35–37, 94,
        131
    clandestine, 36, 84, 123
    covert program, 187
    diffusion method, 33–36, 94
    laser method, 19, 33, 37–38
ETA, 132, 177. See also Basque
    separatist
Ethanol. See renewable energy source
Eurodif, 58
European Bank for Reconstruction
    and Development, 154
European Union (EU), 18, 59, 99,
    141, 143, 155
Evan Mechan Eco-Terrorist
    International Conspiracy
    (EMETIC), 179
exclusion zone, due to radioactive
    contamination, 152
Exelon, 71, 102, 186

Faraday, Michael, 40
fault tree, 141–42
Feiveson, Harold, 131
Fermi, Enrico, 20
fertile material, 11, 38, 205
    plutonium-238, 11, 26
    thorium-232, 11, 24, 38–39
    uranium-238, 11, 24–25,
        30, 33–39, 161, 172, 191,
        205–6
Finland, 65, 74, 198
fissile material, 11, 28, 30, 32–33,
    38–39, 46, 49, 82, 106, 110–12,
    121, 123, 125, 129, 132, 134–36,
    173, 187

fissile (*Continued*)
  americium-241, 11
  plutonium-239, 11, 22–23, 26–27,
    30, 38, 192
  plutonium-241, 26
  uranium-233, 11, 24–25, 38,
    39, 205
  uranium-235, 11, 17, 22–25, 29–30,
    32–39, 42–44, 46, 80, 161, 191,
    205
fission, 5, 7, 10–11, 17, 20–24, 26,
    30–31, 38, 42–43, 46, 49, 80,
    82–84, 93, 110–11, 139–40,
    149–50, 161, 178, 190–91, 193,
    195–96, 205
Fleischmann, Martin, 19
force
  electrical, 8, 10
  gravity, 7–8, 160
  nuclear, strong, 8, 10–11, 17
  nuclear, weak, 11
  tidal, 7
Ford, Gerald, 83
forensic analysis, 16
fossil fuels, 5–6, 16, 55, 60, 62–63, 72,
    87, 91–98, 204
  coal, 5, 15, 16, 54, 56, 60–62, 68,
    69, 70, 91, 92, 94–98, 162, 195,
    197, 206
  natural gas, 5, 56, 58–62, 67–70,
    91–92, 94–96, 206
  oil, 5, 53–56, 60–63, 67, 70, 92, 94,
    148
France, 18, 20, 31, 35, 50, 56–58,
    60, 62–65, 71–72, 80–82, 105, 107,
    109, 111–12, 144–45,
    153, 187
Frisch, Otto, 21
Fuchs, Klaus, 104
fuel assembly, 29, 93, 190, 191
fuel rod, 29
Fukushima Daiichi Nuclear Power
    Plant, accident at, 143, 157, 158,
    163–68, 170–72
Fukushima Daini Nuclear Power
    Plant, 164
fusion, 5–7, 10–11, 17, 18, 22, 24, 204
  cold, 19
  gravitational confinement, 18
  inertial confinement, 18

Gaia hypothesis, 102
Gavin, Francis, 115
Gazprom, 59
General Electric, 76, 102, 160
geothermal. *See* renewable energy
    source
Germany, 20, 22, 57, 59–60, 65, 75,
    104, 113, 152–53, 170
Gigawatt. *See* units of energy
Gilpatric, Roswell, 115
Global Laser Enrichment
    Corporation, 37
Global Nuclear Energy
    Partnership (GNEP), 84–85
global warming, 88–89, 91, 93, 98,
    101
Goldsworthy, Michael, 37
Gorbachev, Mikhail, 152
Government Accountability
    Office, 185
Great Britain, 12, 19, 40, 47, 54, 57,
    101, 105, 109, 177
greenhouse gases, 86, 87, 89, 91,
    93–96, 99
  emissions, 60, 70, 72–73, 87, 91–92,
    95–100, 198
Greenland, 89
Greenpeace, 101, 182, 198
Group of Seven (G-7), industrialized
    nations, 153–54
Gulf War I, 121, 187

Hahn, Otto, 20–21
half-life. *See* Radioactivity
Hall, Ted, 104
Hanauer, Stephen, 167
Harvard University Belfer Center,
    81
Herschel, William, 25
highly enriched uranium (HEU),
    110–11, 124, 134
Hiroshima, atomic bombing of, 104,
    134
Hitachi, 75–76
Human health
  asthma, 198
  cancer, 13–14, 26, 27, 151, 178
  hemorrhaging, 15
  radiation sickness, 14
  respiratory, 99, 198

Hussein, Saddam, 187
hydrogen gas, 165–66

icebreaker, nuclear-powered, 51,
    111
immune system, 15
improvised nuclear device (IND),
    134
India, 18, 39, 44, 50, 57, 64–65, 76,
    80, 100, 103, 106–7, 109, 111–12,
    116, 122, 125, 127–28
Indian Point Nuclear Power Plant,
    181
Indonesia, 67, 161, 178
Institute for Advanced Study, 22
Institute for Nuclear Power
    Operations (INPO), 156–57
International Atomic Energy
    Agency (IAEA), 114, 117–21,
    123–24, 126, 174, 181
International Nuclear Safety
    Advisory Group, 153
International Panel on Climate
    Change, 90
International Panel on Fissile
    Materials, 110, 112
International Thermonuclear
    Experimental Reactor (ITER), 18
iodine. See Periodic Table of the
    Elements
ionizing radiation. See radiation
Iran, 36, 54, 57–58, 67, 106, 117, 119,
    127, 129–30, 133, 187–88
Iraq, 34, 121, 187
Ireland, 115, 177
Israel, 60, 106–7, 109, 116, 122, 127,
    187–88
Italy, 20, 153

Jaczko, Gregory, 165
Japan, 18, 31, 50, 65, 80–82, 84, 99,
    103, 107, 111–12, 117, 120, 133,
    153, 163–70, 179, 180, 206
Japanese Self-Defense Force, 166
Japan Steel Works, 75
Jenkins, Brian Michael, 132
Jiabao, Wen, 162
Johnson, Lyndon Baines, 115–16
Jordan, 67, 79
joule. See units of energy

Juliot-Curie, Frederic and Irene, 20
Jupiter, 27

Kaiser Wilhelm Institute for
    Chemistry, 20
Katsumata, Tsunehisa, 169
Kazakhstan, 80, 114
Kempton, Willett, 207
Kennedy, John F., 114
Kennedy, Joseph, 27
Kepco, 76
Keystone Center, 97
Khan, A. Q., 36–37
kilowatt. See units of energy
kilowatt-hour (kWh). See units of
    energy
Klaproth, Martin, 25
Klein, Dale, 159, 185
Koeberg Nuclear Power Plant, 177
Kristensen, Hans, 110
Kyd, David, 181
Kyoto Protocol, 99, 100

laser enrichment, 19, 33, 37–38
    atomic vapor laser isotope
        separation (AVLIS), 37
    molecular laser isotope
        separation (MLIS), 37
Lash, Jonathan, 102
Lawrence Livermore National
    Laboratory, 19
Lebanon, 133, 182
Levi, Michael, 135
Libya, 36, 64
Lieberman, Joseph, 170
linear no-threshold model, 14
Lithuania, 59, 65, 141, 143, 155
Lovelock, James, 101–2

Manhattan Project, 21–22, 33–34, 104
Markey, Edward, 170
mass, of matter, 7–11, 17–18, 20–21,
    30, 42–43, 190–91
McCain, John, 85
McMillan, Edwin, 27
McNamara, Robert, 115
McVeigh, Timothy, 182
medical isotope, 111
megawatt. See units of energy
Meitner, Lise, 20–21

Meserve, Richard, 183
Mexico, 64–65, 100
Middle East, 67, 106, 187
Milling, of uranium ore, 28–29, 78, 94, 121
millirem, 197
minerals
  autunite, 28
  carnotite, 28
  graphite, 43, 47, 49–50, 150–51, 161
  pitchblende, 25, 28
  tobernite, 28
  uranite, 28
  uranophane, 28
mining
  coal, 162, 197
  strip, 197
  uranium, 28–29, 78, 94, 121
Mitsubishi, 76
Model Additional Protocol to Comprehensive Safeguards Agreements, 121
molecular laser isotope separation (MLIS). See laser enrichment
Moore, Patrick, 100–102
Mosaddegh, Mohammed, 54

Nagasaki, atomic bombing of, 104
National Academies of Science, 195
National Ignition Facility (NIF), 19
National Research Council Nuclear and Radiation Studies Board, 199
negative security assurance, 129
Netherlands, the, 20, 57, 65
neutrinos, 10, 11
neutrons, 8–13, 17, 20, 22–24, 28, 30, 38, 41–44, 46, 49, 51, 139, 149–51, 159, 172
New York Times, 169
Nichols, Terry, 182
Nobel Prize, 21, 27
Non-Proliferation Treaty (NPT), 116–18, 121–23, 126–27
Norris, Robert S., 110
North Korea, 36, 103, 106–7, 109, 112, 116–17, 128, 188
NRG Energy, 71, 102
Nuclear and Industrial Agency, of Japan, 168

Nuclear Energy Agency, 79
Nuclear Energy Institute (NEI), 101
nuclear fuel cycle, 28–29, 32, 38, 76, 84, 94, 131
nuclear fuel market, 57
Nuclear Non-Proliferation Act of 1978, 130
nuclear posture, of the United States' nuclear arsenal, 115
nuclear proliferation, types of
  horizontal, 103, 112
  vertical, 103
Nuclear Safety Account, 154
Nuclear Suppliers Group (NSG), 114, 122, 125
nuclear waste repository, 198
nuclear weapon state, 84, 105, 112, 116, 122–23
  nonnuclear weapon state, 112, 114, 117, 119, 121–22, 126
nucleus, 8–13, 15, 17, 21–24, 26, 30, 42–43

Oak Ridge National Laboratory, 197
Obama, Barack, 85, 118, 170, 200
oil exploration, 54, 70
Okamura, Yukinobu, 168
Oklahoma City, 182
Onogawa Nuclear Power Plant, 164
Organization of Petroleum Exporting Countries (OPEC), 53
Otegi, Arnaldo, 132

Pacala, Stephen, 95–97
Pahlavi, Shah Reza, 54, 57
Pakistan, 36, 65, 103, 106–7, 109, 112, 116, 127, 188
Pebble Bed Modular Reactor (PBMR), 47, 161
Peligot, Eugene-Melchior, 25
Periodic Table of the Elements, 38
  aluminum, 13
  americium, 11, 82, 190
  barium, 20
  carbon, 5, 16, 43, 70, 72–73, 91–92, 94, 96, 97, 100
  curium, 82, 190
  gold, 19
  helium, 6, 9, 12, 15, 24, 49, 161, 204

hydrogen, 5–6, 9, 15–17, 19,
    24, 42–43, 49–50, 63, 105, 128,
    148, 151, 165–66, 204–5
iodine, 166
iron, 9, 24, 40
lead, 13, 49, 88, 162
lithium, 9
neptunium, 30, 190
oxygen, 5, 19, 42, 43, 50, 94, 148,
    151
plutonium, 26, 27, 30–32, 38,
    44, 46, 49, 52, 57, 76, 80–84, 107,
    110–12, 124–30, 134–35, 139,
    171–72, 187–88, 190–92, 196,
    205–6, 209
polonium, 26
protactinium, 38, 39
radium, 26
thorium, 7, 11, 24, 38, 39, 49, 205
uranium, 7, 9, 11–12, 15, 17,
    20, 22–39, 42–44, 46–50, 55,
    57–58, 62, 64, 76, 78–81, 83,
    94, 95, 110–13, 124–25, 127,
    130, 133–35, 139, 161, 171–72,
    187–88, 190–92, 196–97,
    205, 206
xenon, 149, 150
zirconium, 139, 195
petroleum, 5, 16, 62, 91
Philippines, the, 67, 161
phosphorescence, 25
photosynthesis, 6
Pierce, William, 180
plasma, for fusion reaction, 18
Pluto, 27
plutonium. *See* Periodic Table of the
    Elements
Pons, Stanley, 19
positive void coefficient, 150
positron, 12
power plant
    coal-fired, 15, 60, 68, 195, 197, 206
    natural gas, 59, 60–62, 68–70, 92,
        95–96, 206
    nuclear, 137–38, 142–44, 146, 156,
        158–59, 161–62, 175–81, 183–87,
        189, 191, 193, 195, 197, 200,
        206–7
probabilistic risk assessment (PRA),
    141

Project on Government Oversight, 186
protons, 8–9, 11, 13, 17, 24, 26, 42
PUREX, type of reprocessing
    technology, 83–84

radiation
    alpha, 12–13, 27, 39, 171, 192
    beta, 12–13, 15, 192–93
    detector, 16, 135, 136
    gamma, 10–11, 13, 27, 51, 192
    ionizing, 11, 13–15, 151, 157, 176,
        189, 192
radioactive isotopes
    cesium-137, 15, 26, 177, 192
    cobalt-60, 15, 26
    plutonium-238, 11, 26
    strontium-90, 192–93
    tritium, 15, 17–18, 128
Radioactivity
    decay, 7, 11, 15, 16, 30, 31, 38, 140,
        160, 189–92
    in fly ash, 197
    half-life, 15, 16, 24–26, 38, 39, 192
    radioisotope, 12, 15, 16, 26, 51, 52
Radkowsky, Alvin, 39
Ramberg, Bennett, 186
reactor
    AP-1000, 160
    boiling-water reactor (BWR), 45
    boiling water reactor, Mark I,
        167–68
    breeder, 49, 205, 206
    CANDU (Canadian Deuterium
        Uranium), 44
    CIRUS (Canadian Indian
        Research United States), 125
    Economic Simplified Boiling
        Water Reactor (ESBWR), 160
    evolutionary (or European)
        pressurized reactor, 73, 160
    gas-cooled fast reactor, 49
    Generation I, 48
    Generation II, 48–49
    Generation III, 48
    Generation IV, 48, 50
    lead-cooled fast reactor, 49
    molten salt fast reactor, 50
    Monju fast reactor, 50, 82
    Pebble Bed Modular Reactor
        (PBMR), 47, 161

reactor (*Continued*)
pressurized-water reactor (PWR),
44–45, 48
Radkowsky Thorium Reactor, 39
sodium-cooled fast reactor, 50
Soviet designed, 59, 153–55
supercritical water-cooled
reactor, 48–49
Superphenix, 50
Tarapur, 127
VVER-440, 154–56
VVER-1000, 155–56
reactor-grade, type of plutonium,
32, 110
reactor meltdown, 47
reforestation, 93
regulatory agencies, 67, 162, 182
Reiss, Mitchell, 113
renewable energy source, 7, 60, 204–5
biofuels, 6, 204–5
ethanol, 54–56, 97, 204
geothermal, 7, 68
methanol, 56
solar, 5–7, 56, 60, 62, 73, 91, 204–5,
207
wind, 6–7, 56, 60, 62, 73, 91, 204,
207
reprocessing, of spent nuclear fuel,
30, 32, 81, 83–85, 125
clandestine, 36, 84, 123, 187
multiple recycling, 31
PUREX, 83–84
single recycling, 31–32, 81
research and development, 73, 93,
105, 159
Richardson, Bill, 128
Rickover, Hyman, 47
Rixin, Kang, 162
Roosevelt, Franklin Delano, 22
Rosatom State Nuclear Energy
Corporation, 76
Russia, 18, 31, 43, 50, 51, 57–60, 65, 71,
76, 80, 99, 103, 107–12, 114, 127,
128, 133, 140–42, 156, 177, 180
Rutherford, Ernest, 12, 20

sabotage, 174–75, 181, 199
airplane crash, 68, 181, 184, 186
commando attack, 181–83
cyber attack, 181, 184

by insiders, 183
truck bomb, 133, 181–82
waterborne attack, 181
safety
active, 156, 159–61
advanced passive, 160
culture, 67, 137–38, 142, 153, 158,
175
passive, 142, 159, 160–61
sarcophagus, for damaged
Chernobyl reactor, 153
sarin gas, 179
saturation threshold, 90
Saturn, 27
Saudi Arabia, 64, 67
Science, 95
Seaborg, Glenn, 27
September 11, 2001, 119, 133, 176,
181–85
Sheffield Forgemasters, 75
shielding, for radiation protection,
190–91, 193
Siberia, 89, 129
Silex, laser enrichment technique,
37–38
Slovakia, 56, 58–59, 65, 143,
155
Socolow, Robert, 95–96
Sofidif, 58
Sokolski, Henry, 123
solar. *See* renewable energy source
solar photovoltaic panel, 6
South Africa, 57, 64–65, 113, 177
South Korea, 18, 44, 66, 74, 76, 103,
107, 180
special theory of relativity, 21
speed of light, 9, 13, 21
spent nuclear fuel, 26, 29–32, 44,
51, 81–82, 112, 124–25, 190–96,
199–200, 202, 206
stable isotope, deuterium, 17, 18,
43, 205
Statoil, 92
steam generator, 40, 45, 49, 75, 141,
147, 154
stockpile, 80, 109, 135
civilian, 23, 110–13, 131
military, 110, 112
separated plutonium, 110,
112

storage, 29, 51, 81, 173, 189, 197, 199–202, 207
dry storage cask, 30, 191, 193–94, 196
pools, for spent nuclear fuel, 30–31, 191, 193–96
Strassmann, Fritz, 20
Strategic Arms Reduction Treaty (START), 108
strategic weapon system, 108
intercontinental ballistic missiles (ICBM), 108
long-range bomber, 108
nonstrategic weapon system, 108
submarine launched ballistic missile (SLBM), 108
Study of Terrorism and the Responses to Terrorism, 180
submarine, 47, 51, 105, 108, 110
supernova, 24–25
supply chain, 74–75
sustainability, of energy system, 203, 205
Sweden, 20, 65, 152, 198
Syria, 188
Szilard, Leo, 22

Taiwan, 63, 66, 105, 180
Tarapur Power Reactor, 127
Teller, Edward, 22
Tennessee Valley Authority, 128
Thailand, 67
thermal diffusion, of uranium enrichment, 33–34
thermal neutrons, 42–43
thermal reactor, 42, 46, 50, 82
thermonuclear reaction, 19
thorium fuel cycle, 38–39
Three Mile Island Nuclear Power Plant, 61, 138, 140, 143, 146–48, 156–58, 167
tipping point, 91, 106, 152
Tokai Nuclear Power Plant, 164
Tokyo Electric Power Company (TEPCO), 168, 169
Toshiba, 75–76
toxic waste, 194–95
transformation, of radioactive materials, 4, 11
Transportation Research Board, 199
Truman, Harry S., 105

tsunami, hazard to nuclear facilities, 163
turbine, 6, 40, 45–46, 49, 51, 62, 147, 150, 207
Turkey, 58, 64, 161
Turner Diaries, The, 180

Ukraine, 56, 58–59, 66, 114, 149, 152, 153, 180
Unistar, 72
United Arab Emirates (UAE), 66–67, 70, 131
United Nations, 99, 114–15, 118
Framework Convention on Climate Change, 99
IAEA (See International Atomic Energy Agency)
Security Council, 114, 119
United Stated Climate Action Partnership (USCAP), 102
United States, 18, 22, 34–35, 41, 45, 47, 54–57, 60–62, 64, 66, 69, 70, 73, 77, 79, 83–85, 96, 99–100, 102–12, 115–20, 123, 125–31, 133, 139, 143, 145, 148, 153, 155, 159, 179–80, 183, 189, 193, 197, 199–201, 203
United States Enrichment Corporation (USEC), 94
units of energy, 4
electron volts, 16, 17
gigawatt (GW), 4, 96, 162
joule, 4
kilowatt, 4, 5, 69, 74, 191
kilowatt-hour (kWh), 5, 73, 95, 201
megawatt, 4, 41, 50–51, 62, 73, 97
watt, 4, 95
unstable, 7–8, 10–13, 38, 103, 140, 153
uranium. See Periodic Table of the Elements
Uranus, 25, 27
Urenco, 57
U.S. Atomic Energy Act, 72
U.S. Department of Energy, 76, 200
U.S. Energy Information Administration, 61
U.S. Environmental Protection Agency, 101

U.S.-India nuclear deal, 39, 50, 57, 80, 122, 127
U.S. Navy, 48, 145, 156
  propulsion, of warships, 39, 51, 111
  submarine reactor program, 47
U.S. Nuclear Regulatory Commission, 77, 143, 148, 156, 160, 165, 168, 182, 189, 201

Venus, 87–88
Vietnam, 67
von Hippel, Frank, 168

Wahl, Arthur, 27
Washington Post, 168
waste heat, 40, 98
waste management
  bentonite clay, 194
  drip shield, 194
  repository, 82, 194, 198–202
  vitrified waste, 194–95
water
  heavy water, 19, 43–44, 125
  light water, 43–44, 145, 190
  spray system, 196
water vapor, 44, 86, 88
watt. *See* units of energy
weapons
  biological, 179–80
  chemical, 179

nuclear, 11, 16, 22–23, 26, 28, 32–33, 38–39, 104–6, 110, 113, 121, 173, 179–80, 187
  radiological, 16, 177, 180
West Germany, 113
Westinghouse, 75–76, 160
Whitman, Christine Todd, 101
Wigner, Eugene, 22
Wilkinson, Rodney, 177
wind. *See* renewable energy source
wind farms, 60, 207
Wolfson, Richard, 4
World Association of Nuclear Operators (WANO), 157
World Nuclear Association, 64, 195
World Resources Institute, 102
World Trade Center, 176, 181–82
World War I, 54
World War II, 34, 104–5

yellowcake, uranium ore concentrate, 28
Yemen, 67
Yousef, Ramzi, 182
Yucca Mountain, 200–201

Zanger Committee, 126
Zedillo, Ernesto, 124
Zedong, Mao, 105